MERCURY
RETROGRADE

MERCURY
RETROGRADE

Its Myth AND *Meaning*

Pythia Peay

Jeremy P. Tarcher/Penguin

a member of Penguin Group (USA) Inc.   New York

Most Tarcher/Penguin books are available at special quantity discounts for bulk purchases for sales promotions, premiums, fund-raising, and educational needs. Special books or book excerpts also can be created to fit specific needs. For details, write Penguin Group (USA) Inc. Special Markets, 375 Hudson Street, New York, NY 10014.

While the author has made every effort to provide accurate telephone numbers and Internet addresses at the time of publication, neither the publisher nor the author assumes any responsibility for errors, or for changes that occur after publication.

Jeremy P. Tarcher/Penguin
a member of
Penguin Group (USA) Inc.
375 Hudson Street
New York, NY 10014
www.penguin.com

LIBRARY OF CONGRESS CATALOGING-IN-PUBLICATION DATA

Peay, Pythia.
Mercury retrograde : its myth and meaning / Pythia Peay.
p.   cm.
Includes bibliographical references and index.
ISBN 1-58542-316-5
1. Astrology.   2. Mercury (Planet)—Miscellanea.   I. Title.
BF1724.2.M45P43   2004   2004040757
133.5'33—dc22

Printed in the United States of America

1   3   5   7   9   10   8   6   4   2

Book design by Kate Nichols

ITHACA

When you set out on your journey to Ithaca,
pray that the road is long,
full of adventure, full of knowledge.
The Lestrygonians and the Cyclops,
the angry Poseidon—do not fear them:
You will never find such as these on your path,
if your thoughts remain lofty, if a fine
emotion touches your spirit and your body.
The Lestrygonians and the Cyclops,
the fierce Poseidon you will never encounter,
if you do not carry them within your soul,
if your soul does not set them up before you.

Pray that the road is long.
That the summer mornings are many, when,
with such pleasure, with such joy
you will enter ports seen for the first time;
stop at Phoenician markets,
and purchase fine merchandise,
mother-of-pearl and coral, amber and ebony,
and sensual perfumes of all kinds,
as many sensual perfumes as you can;
visit many Egyptian cities,
to learn and learn from scholars.

Always keep Ithaca in your mind.
To arrive there is your ultimate goal.
But do not hurry the voyage at all.
It is better to let it last for many years;
and to anchor at the island when you are old,
rich with all you have gained on the way,
not expecting that Ithaca will offer you riches.

Ithaca has given you the beautiful voyage.
Without her you would have never set out on the road.
She has nothing more to give you.

And if you find her poor, Ithaca has not deceived you.
Wise as you have become, with so much experience,
you must already have understood what Ithaca means.

—CONSTANTINE P. CAVAFY, 1863–1933

CONTENTS

"Pluto is squaring my Mars!" And they would get it—understand that contained in that cryptic statement were thousands of years of myth, symbol, human experience, art, music, sculpture, and literature that magically transmuted events in my life from being a random series of events to a charted encounter with a meaningful fate.

Aristotle said that a person does not really know a subject until he can teach it. With this book I have endeavored to offer the fruits of my own years-long studies in astrology, which, in turn have been deepened and enhanced by my work as a writer and teacher on spiritual themes. But teachers are students first, and perhaps always. By cultural definition, astrologers are an eccentric group of independent, individualistic thinkers. They function outside the mainstream, where they are free to let their intellects and imaginations roam, combining history, politics, astronomy, and myth in deft and intriguing ways that yield insights not found anywhere else. And so I would like to thank those astrologers who, in person and through books and articles, have helped me decode the mysteries of astrology. First, I would like to thank those astrologers who so generously gave of their knowledge in interviews for this book: Lynn Koiner, whose lively classes first helped to bring the stars alive in a practical and accessible way; Dana Gerhardt, my soul sister and partner at MoonCircles, whose astrological wisdom and writerly imagination are joys to share, and whose guidance on this book has been invaluable; Erin Sullivan, for her scholarly commitment to astrology, and for mentoring me in the mysteries of the planetary retrograde cycles; Laurence Hillman, for his magical symbolic imagination, and for his lovely friendship and insights into my own natal chart; Ray Grasse, for bringing astrology into the twenty-first century with fresh perspectives; Robert Schmidt, for so generously sharing the fruits of his historical research into astrology's classical foundations, and his partner Ellen Black; Edith Hathaway, for her sharp insights into how astrology affects our daily lives; Chakrapani Ullal, for his blend of spiritual wisdom and hands-on astrology; and April Kent, who so generously answered my panicked call for astrological computer help. Many thanks to writer

Narelle Bouthillier for putting into words her colorful escapades with Mercury retrograde for use in this book. I would also like to thank those friends with whom I have shared stimulating exchanges on astrology over the years: Susan Roberts, for her insights into the planetary archetypes, and for sharing with me an enthusiasm for myth, dreams, and the meaning of life; Sylvia Seret, for her knowledgeable input into this book, and for our near-daily phone conversations as we observe the influence of the stars on the lives of our children, ourselves, and the world; Dodie Brady, who shares my passion for how astrology impacts world events; astrologer Jean Lall, whose original perceptions always reveal a new dimension of understanding regarding a particular planet or transit; and astrologer Caroline Casey, whose year-in, year-out celebration of the solstices and equinoxes, along with her revolutionary running commentary on the political gestalt, has enriched my life immeasurably; and I would like to thank astronomer Sean O'Brien at the National Air and Space Museum, for helping me to understand the astronomical underpinnings of the Mercury retrograde cycle.

I would like to thank those astrologers, both living and departed, for the numerous books, magazine articles, and recorded lectures they have produced: The Jungian analyst and noted astrologer, Liz Greene, whose books were pivotal in my education, as well as Howard Sasportas, Robert Hand, Stephen Arroyo, Arielle Guttman, Richard Tarnas, Demetra George, Lois Rodden, Donna Cunningham, Dale O'Brien, Richard Idemon, Charles Harvey, Ray Grasse, Isabel Hickey, Charles Carter, Steven Forrest, Elias Lonsdale, Dane Rudhyar, Maya Del Mar, Noel Tyl, Jeff Greene, Martin Shulman, Martha Pottenger, Tracy Marks, Dennis Harness, Edith Hathaway, and countless others too numerous to mention in one place. I would like to thank Tem Tarriktar of the *Mountain Astrologer,* for publishing an in-depth magazine on contemporary astrological trends, as well as Mark Lerner, publisher of *Welcome to Planet Earth.* Thanks also to Jeff Jawyer and Rick Levine of stariq.com, for their provocative and informative website that begins each day.

As always, I would like to thank my network of friends for laughter and support: Lis and Jerry, Elise, Yvonne and Rich, Jessica and Richard, Harriett, Dodie and Jim, Barbara and Hugh, Debbie, Kristen, Jack and Pauline, Azimat, Liz, Blanchefleur, Anne, Reza, Zuleikha, Tatiana, Ann, Tara, Amida, Zarifa and Munawir, Sharon, Corinne and Gordon, and especially Taj Inayat, for her golden heart and generous spirit, and Yvonne Seng, for chocolate and laughter. I would like to thank Carol Montoni of CM Media for so happily giving of her creative computer skills and designing my beautiful website, my brother-in-law, Tim Peay, for his tech support, as well as Susan Rodberg, for once again patiently helping to heal my computer "crises." Great personal thanks are due to my editor, Mitch Horowitz, for his passion for astrology, commitment to the world's wisdom traditions, and for holding my feet to the fire and encouraging me to write, write again—and then again; thanks also to his assistant, Ashley Shelby. To my publisher at Tarcher, Joel Fotinos, for his inspiring friendship and support; to publicist Ken Siman for his bouyant enthusiasm for this book; Nancy Ellis for her continuing encouragement; and Liz Williams for her loving spirit, savvy head, and for years of sharing in the excitement of those big ideas we both care so much about. To Grace Ogden and the beautiful women on the National Cathedral Sacred Circles committee, and to my Sufi community for the sacred friendship we have shared over the years.

And last, I would like to thank my family for giving me heart and hearth-warmth: Terry and Anne Peay, for friendship and for sharing in the ongoing adventure of raising our three sons. Carol Montoni and Alison Forestal, for their intelligent friendship asnd loving support. To my brother Steve, for his quiet but generous support, and to John and Tracy, Colleen and Z, and my mom for the richness of family. And to my sons Kabir, Amir, and Abe who, as they grow older, have become dear friends and lively thinkers in their own right: You are my heart's treasures who make life and work worthwhile.

PART ONE

The Myth

1.

MERCURY RETROGRADE: A HEALING MYTH FOR OUR TIME

We shall not cease from exploration
And the end of all our exploring
Will be to arrive where we started
And know the place for the first time.

—T. S. Eliot, "Little Gidding,"
(the last of his *Four Quartets*)

A woman who has just departed on a long-awaited tour to Sicily finds herself trapped overnight in an airport in Chicago, delayed by an unexpected snowstorm. Finally arriving at her destination, she discovers that her luggage is missing.

A journalist submits an article to her editor by e-mail. But after he edits and returns it to her by electronic transmission, the story is mysteriously missing from the body of the e-mail text. Each time he resends the article, nothing comes through. After several failed attempts, he finally gives up and faxes her the article for approval.

His checking account on empty, a man drives hurriedly to make a deposit before the bank closes. Pulling into the bank parking lot, he finds himself at the tail end of a long line of stalled traffic, blocked by a delivery truck that has broken down in the driveway. Later, he discovers that the check he was finally able to deposit on time didn't clear—because the person who wrote it mistakenly grabbed a checkbook from an account that had been closed.

A woman who has spent months developing an idea for a book travels to a writer's conference, where she looks forward to meeting with a literary agent. There, she is told by the agent that her idea is vague and uninteresting, and that she should find a different topic. Discouraged, she returns home, making up her mind to stop writing altogether.

Despite their modern-day guise, these stories illustrate the age-old garden-variety delays, doubts, and setbacks that bedevil us all at one time or another. As familiar a part of life as they may be, most of us never cease being caught off guard by their unwanted intrusion. Rarely granted forewarning of these petty grievances, it is easy to become ensnared in tension and frustration.

Astrologers, however, know that three or four times a year, for a period of twenty-one days, there is a greater likelihood that our plans may be thwarted. They call this time period Mercury retrograde. This is the thrice-yearly astrological event that is said to throw schedules off-kilter, derail the best-laid plans, cast doubt on cherished dreams, stir misunderstandings and doubt in personal relationships, and make computers malfunction and cars break down. Under its influence the daily commerce of everyday life slows to a crawl and becomes backed up, tied up, and mired in confusion and anxiety.

Chances are, you may have already heard of it. For increasingly, even the most skeptical are debating the effects of this mysterious-sounding phenomenon that could be said to be the ghost in the machine of our efficient, smooth-running society.

Indeed, Mercury retrograde's legendary ability to stir confusion and chaos in its wake is generating widespread curiosity. Many in business, technology, and the media want to learn more about avoiding the pitfalls and obstacles that often attend this cycle. For while most people's knowledge of astrology doesn't extend beyond reading their Sun sign forecast in the daily newspaper, growing numbers now take serious notice of this triannual event.

Along with the new and full Moons, some calendars now mark the period of time during a particular month when Mercury will be retrograde. Whether at parties or around the water cooler, at yoga class or waiting in line at the grocery store, it is not unusual to hear tales of the latest Mercury retrograde "horror stories." Even a September 30th, 1996, column in the *New York Times* described the "fear and trembling" that arises in Internet circles during the planet's apparent reversal in the sky. "Technology buffs who do subscribe to these principles are legion," writes journalist Edward Rothstein, "and the World Wide Web sites, Usenet groups and on-line astrology services have been offering warnings for weeks about the period of Mercury's strange motion in the heavens in which it seems to change direction and back up."

But what exactly *is* Mercury retrograde—those cycles that cause even nonastrologers to cancel plans, stall important decisions, and put life on hold?

Three to four times a year the planet Mercury appears to slow down and reverse direction. Like a car that shifts into reverse and moves backward over a stretch of road it has just covered, the planet seems to retrace a section of its pathway around the Sun. It then shifts forward, moving "direct" again until it returns to its original starting point. At this stage it begins to move more quickly along its orbit until several months later, when it repeats the entire process all over again. Each period of time Mercury is retrograde lasts approximately three weeks.

Mercury does not actually stop in the heavens and go backward, of course—it just appears that way to those on Earth who are observing its motions. This celestial optical illusion occurs whenever Mercury, the nearest planet to the Sun, begins to round the bend in its orbit and swing closer to Earth. Thus, from Earth's position, Mercury appears to be moving backward against the sky. The simple example of a racetrack, whose curvature is similar to the orbit of the planets around the Sun, helps to explain this phenomenon. Imagine, for instance, that you are a passenger traveling in a car around a racetrack. Soon, a car traveling on an inside track, but at a greater speed, begins to catch up to you. As it reaches a parallel position to your car, it momentarily appears to be moving backward against the stationary backdrop of buildings, before finally moving ahead. The retrograde phenomenon is not unique to Mercury, but affects all the other known planets in our solar system, including Venus, Mars, Jupiter, Saturn, Neptune, Uranus, and Pluto.

So why should a person in the twenty-first century believe that an optical illusion in the skies above influences our activities on earth below? No one really knows how astrology "works." But the simplest explanation of astrology is that it is the symbolic interpretation of the planetary movements in our solar system. As they have done for millennia, astrologers read the stars for signs in a way that imparts meaning to human affairs.

Ancient astrologers believed that the cosmos was an intricate and interrelated creation, known variously as the doctrine of correspondences or "sympathies." In this view the planets and their cycles represent potent mythic forces that mysteriously reflect the human condition, allowing us a glimpse behind the curtain that separates the seen and unseen worlds. Today, as centuries ago, astrology rests on the belief that the universe is animated with meaning—and that a guiding spirit of intelligence links the microcosm of mundane affairs to the larger macrocosm.

The practical work of astrologers involves calculating and interpreting birth charts for individuals, as well as for larger entities, such as corporations or nations. These charts are based on the positions of the planets at a specific moment in time. But astrologers also track the movements of planets in order to predict widespread cultural trends that affect society in general. In this way, they can offer helpful insights to people, providing them with a kind of astrological "weather forecast."

Among other predictive techniques, astrologers pay close attention to the periods when any one of the eight planets are retrograde. As astrologer Rick Levine explains in "Mercury Retrograde: A Modern Look," a November 1999 article he wrote for his astrology Web site stariq.com, a planet always retrogrades when it is closer to the Earth. This close proximity of a planet's orbit to Earth, writes Levine, makes it "louder" than when it is far away. Thus a retrograde planet, says Levine, "is actually more powerful than when it is moving in its normal direct motion."

Because Mercury retrograde cycles are shorter and occur more frequently, however, astrologers say their effects are more pronounced than those of the other planets. And because astrological tradition gives the planet Mercury "rulership" over the sphere of life that includes commerce, trade, transportation, and communication—activities that affect virtually everyone—its influence is even more enhanced. Like the stick in the spoke of the wheel that throws a cart off course, Mercury's reversal can often mean setbacks, obstructions, misunderstandings, and sudden, unexplained "out-of-the-blue" events.

For these reasons, astrologers traditionally counsel their clients to handle their mundane affairs with greater caution and increased awareness during the tricky period of Mercury retrograde. Clients are urged to delay signing contracts, to put off buying new items—especially anything mechanical or technological—and to hold off initiating new ventures. They are warned that any plans made under Mercury retrograde can be subject to revision. Personal relationships or professional projects that are under way are likely to take surprising twists and turns; people may become doubtful, change their minds, and suddenly reverse course. Likewise, it is a time to check details carefully whenever ordering or sending anything on line or through the mail. While Mercury retrograde is considered by some astrologers to be a favorable time to go on vacation, there can be problems with transportation delays or lost luggage.

Yet for all the confusion that surrounds Mercury retrograde, it contains within it a vein of gold—if only we would mine it for its true worth. By working *with* the slower pace of Mercury retrograde, rather than going against it, we can allow ourselves to step outside the narrow framework of ordinary reality and see our lives from a wider perspective. Most astrologers, for instance, agree that it is a cosmically timed period to undertake any activity with the prefix *re:* redo, reevaluate, revise, rethink, review, reconsider, reassess. Whether restoring an antique, reordering a messy closet, rewriting a business proposal or article, reflecting upon a relationship, revisiting old friends, or taking time out to hike or laugh over a glass of wine, Mercury retrograde is all about loosening the grip of everyday demands and responsibilities in order to experience life more deeply. It is about finding pleasure in the old, familiar, and timeless, rather than the new and untried.

A dream I had during an especially frustrating Mercury retrograde period reveals the kind of attitude we need to cultivate during this time. In it, I had enrolled in college. It was the first day of school, and I decided to park my car and walk with the students streaming onto the campus. Unlike me, the students were all young, riding skateboards or bikes, colorfully dressed in unconventional clothing, and brimming with enthusiasm. When we reached the doors to the lecture hall, however, we were greeted by a note from the professor saying that he would be late to class. Suddenly it began to rain, soaking our clothes and books. But rather than complain, the crowd of students began to laugh and dance. Throwing my hands up to the sky, I joined in, moving joyously in rhythm with the dancing bodies around me. As the message in this dream imparts, there are times in life when the wisest course of action is to simply let go and savor things just as they are. Thus, rather than push forward, it is a time to *retreat,* if possible, and step back from the usual hurried, harried way of doing things. It is a time to *reflect* in solitude on what has gone on in the months preceding Mercury retrograde, and to reconnect with one's soul. Following these steps makes it possible to emerge from the period of Mercury retrograde refreshed and *renewed* in spirit.

ENTERING A MYTHIC TIME ZONE

In the ancient world, the gods and goddesses embodied a particular essence or represented an important area of life. Each of these deities then became associated with a planet. Thus astrology, based on the movements of the planets, is a kind of theater of living mythology in action. Mars, for example, exemplifies courage and represents any kind of risk taking or martial venture. Saturn represents the principles of discipline, structure, and limitation. Mercury, on the other hand, was frequently depicted as the winged "messenger god" who shuttled between heaven and earth, delivering messages from the gods to the humans. For this reason, astrologers saw the planet Mercury as symbolic of the transmission of knowledge and ideas.

Because we live in an information age, it could be said that of all the planets Mercury and its symbols, cycles, and myths has increased significance for the times we live in. Many astrologers have noted the growing influence of the planet Mercury over modern-day life, depicting it as the "ruling god" of our culture. In his July 2000 article for stariq.com, "Spicing Up Your Mercury Retrograde," astrologer Eric Francis calls Mercury the "cosmic modem." Indeed, the present-day world, writes astrologer Elias Lonsdale in his book *Inside Planets*, "is enraptured with Mercury. This planetary energy shows up everywhere—humor, movement, variety, and restless currents give Mercury access. This is the god of a consumer culture, the way of life of those who are drawn to the surface to find gratification."

If Mercury is the ruling god of our consumer-driven, communications society, however, it is sometimes a god run amok—a distorted, frenetic Mercury that is symptomatic of how our natural rhythms have been thrown dramatically off-kilter and out of balance. While there are benefits to the increased openness and exchange the world enjoys as a result of global communications, few could doubt that the atmosphere we live in has become so mentally overstimulated by "information overload" that many of us cannot sleep well at night. Anxiety disorders abound, deep relaxation is a rare experience, and lack of time is considered a leading source of stress.

One of the purposes of astrology is to mark the natural variations in the rhythms of time. In this sense, the retrograde periods of Mercury offer a built-in opportunity for a retreat or a sabbatical from the usual nonstop way of living, and to develop the more thoughtful, contemplative side embodied in the myths of Mercury. One reason for the current fascination with Mercury retrograde, in fact, may be the potential healing message this cycle could bring to our time-bound, media-driven society. If we paid attention to its message, Mercury retrograde might satisfy an important lack in the modern soul—the need for more slow, unstructured, thinking time.

There is great foresight, for instance, in the cautionary advice many astrologers issue around the retrograde cycles. To deliberate over our actions or step back from our affairs helps return us to a saner, more measured pace of life. For while it may seem outwardly as if things are going wrong and life is falling apart, these retrograde cycles appear differently when viewed from an interior perspective. If we slow down with Mercury's regressive motion, retracing our steps as it retraces its path along the zodiac, for instance, we might discover treasures that we had overlooked in our previous haste—such as stillness, patience, quietude, insight, and reflection.

By stepping outside the onrushing stream of life and following Mercury's backward-moving steps, we cross over into an alternate realm that is more mysterious and multilayered. For in resisting the relentless drive forward that so insistently pushes and pulls at all of us, we invite the inner world to enter our overcrowded consciousness. Working within the cycle of Mercury retrograde this way unlocks the door to the secret mysteries of time—mysteries our own modern-day culture has ignored at the expense of our psychological, spiritual, and physical health.

Like buried archaeological ruins, an ancient template lies beneath the monotheistic traditions of today's religions—an invisible structure that once organized the world into sacred and profane time. This template spread across the Old World cultures like a calendrical cosmic grid; solstices and equinoxes, the phases of the Moon, and the rising and setting of the stars and planets determined religious festivals and ceremonies, rhythmically synchronizing with nature the lives of various tribes and cities across the planet. Though long-forgotten, this archaic memory lives within each of us, stirred awake by the sight of the stars above.

Taking three-week "time-outs" several times a year to realign our souls with something more time-sacred and transcendent than our ordinary concerns may be one way to plug back into this ancient mythic template. By honoring Mercury retrograde, we can begin to mine this extraterrestrial terrain that lies buried beneath the tangled jungle of modern-day life. The psychologist Carl Jung, for instance, often spoke of the "two-million-year-old man" within each of our psyches. And neuroscientists have documented the force of the instincts that lie coiled in the stem of our brains like a sleeping serpent. Thus while Mercury direct favors the logical, rational, left-brained mind, Mercury retrograde awakens our wiser, more intuitive, imagistic right-brained way of thinking and perceiving. Not permanently— the astrological worldview is always fluid and changing—but for a period of time.

Most of the major religious traditions, for example, have encouraged students and initiates to undergo periodic retreats for the purpose of spiritual renewal. While not everyone can go off on retreat three times a year during Mercury retrograde, they can, however, consciously work with their attitudes to live life differently—scheduling more downtime, going away for the weekend, decreasing workloads, sleeping more, experiencing the arts, meditating, journaling, and doing dream work. Whether it's something we've said that has unintentionally misfired, or a storm that causes a power failure, the purpose of Mercury retrograde is to throw us back on ourselves—to redirect our attention inward.

Just as important, Mercury retrograde is a time to lighten up and not take the duties of life so *seriously*. Over the many years of observing Mercury retrograde, one of the things I have noticed is that it is a perfect opportunity to give ourselves a break from trying so hard to achieve our goals and get ahead. For whether or not we do this on our own, Mercury is likely to come along and, in his usual trickster fashion, find a way to upend our neatly arranged lives for us. Why does the impish god do this? To prompt us off our well-beaten paths, I believe, onto life's unexplored byways.

Indeed, in classical mythology Mercury is often referred to as the patron of travelers. As depicted by his winged cap and traveling staff, this ancient pagan god delivers the valuable message that life is a journey—a never-ending adventure of unexpected twists, unpredictable turns, and delightful surprises. As we leave youth and assume the responsibilities of adulthood, however, memory dims and we forget what it was like to experience life as freshly minted and open-ended. Bogged down in our routine, we miss the trip that we are on. Oppressed by meeting all the expectations demanded of us, we leave unopened the divinely inspired gift that life really is. With Mercury retrograde as our guide, however, we can learn to slow down enough to savor the miracles that lie in wait for us along the long highway of life. "When you set out on your journey to Ithaca, / pray that the road is long, / full of adventure, full of knowledge," wrote the poet Constantine P. Cavafy in 1911, ". . . do not hurry the voyage at all. / It is better to let it last for many years; / and to anchor at the island when you are old, / rich with all you have gained on the way . . ."

THE PATH OF MERCURY RETROGRADE AS A SPIRITUAL PRACTICE

The central aim of this book is to gently initiate the reader into a deeper understanding of Mercury retrograde. Through astrological techniques, mythological images, and spiritual perspectives, it will show readers how to work with the retrograde periods as a way to shift from mundane to sacred time, and to follow the phases as a path of introspection and creative contemplation.

A genuine appreciation of Mercury retrograde as a period of deep restoration, however, cannot be gained without an understanding of the astrological tradition within which it is embedded. For this reason, chapter 2 provides a historical overview of astrology, outlining in broad strokes its evolution as one of humankind's oldest belief systems. Chapter 3 explores the classical myths that surround Mercury, called Hermes by the Greeks, as the astrology of Mercury stems from the ancient stories that have grown up around this god over the centuries. With these mythological understandings as a base, chapter 4 then explores the planet Mercury from an astrological perspective and examines how it operates in individual charts. Next, chapter 5 delves into a more specific explanation of the astronomy behind Mercury's retrograde cycles, as well as an astrological interpretation of how Mercury retrograde influences us in a personal and individual way.

The second section of the book, "The Path," outlines a step-by-step approach showing how Mercury retrograde can be used as a spiritual practice. Drawing upon the history of Mercury as he appears in myth, literature, psychology, and astrology, I have explored each of the phases of the retrograde cycle as it exemplifies one of the god's many characteristics. Throughout this section I have also drawn upon Homer's epic journey saga, *The Odyssey,* as a mythic mirror for understanding the wisdom that can be gained from those times in our lives when we are lost, confused, and blown off course.

Chapter 6 addresses the "threshold struggle" that can arise when Mercury first begins to move backward and the security of our daily routines is upset. It is at this initial juncture that Mercury mischievously launches the retrograde phase with its trademark frustrations and disturbances. While this chapter addresses many of the practical concerns raised by those who worry over how to handle their mundane affairs while the planet is moving backward, it also reflects philosophically upon the obstacles that fall across our path as mystical gifts in disguise. From there, chapter 7 moves with Mercury as inner teacher into the stillness of silence, where we turn from extroversion to introspective moral reflection. Chapter 8 awakens the "muse of memory," leading us down into the labyrinths of the unconscious and the realms of the underworld, while chapter 9 shows how Mer-

2.

ASTROLOGY: A SHORT HISTORY OF A PERENNIAL PHILOSOPHY

There can perhaps be no more striking proof of the power and popularity of astrological beliefs than the influence which they have exercised over popular language. . . . Do we still remember, when we speak of a martial, jovial, or lunatic character, that it must have been formed by Mars, Jupiter, or the Moon, that an *influence* is the effect of a fluid emitted by celestial bodies, that it is one of these "*astra*" which, if hostile, will cause me a *disaster,* and that, finally, if I have the good fortune to find myself among you, I certainly owe it to my *lucky star?*

—Franz Cumont, *Astrology and Religion Among the Greeks and Romans*

It then occurred to me that outer space is within us inasmuch as the laws of space are within us; outer and inner space are the same.

—Joseph Campbell, *The Inner Reaches of Outer Space*

Before there were books, before there were buildings, before there was even fire, perhaps, there were the starry sky and human beings gazing up in wonder.

Like a dark and fathomless divining bowl, the night sky has always drawn out the questions hidden within our souls. Written in the glowing letters of a living scripture, the cosmos was the first holy text; in its pages were sought answers to the mystery of the human condition. From humanity's distant ancestor on the African delta to the computer expert in his cubicle, something deep within us has always stirred in recognition whenever we lift our gaze into the beyond. As if touched by a memory of our birth in a long-ago and faraway place, we seek to remember, to recall the truths of our origins in the ancestral womb of space.

"We know," wrote mythologist Joseph Campbell in *The Inner Reaches of Outer Space,* "that we have actually been born from space, since it was out of primordial space that the galaxy took form, of which our life-giving sun is a member. And this earth, of whose material we are made, is a flying satellite of that sun . . . Our eyes are the eyes of this earth; our knowledge is the earth's knowledge. And the earth, as we now know, is a production of space."

The study of the stars was the source of inspiration for humankind's earliest nature-based religions, the foundation myth for much that was to come later. The god of the Sun and the goddess of the Moon divided the days into months and seasons; their cycles created the sacred round of ritual. In some premodern civilizations, each day was sacred to a planet, and prayers were addressed to these stellar deities. Traces of this astral religion are faintly reflected in the Bible: "And God said, Let there be lights in the firmament of heaven to divide the day from the night; and let them be for signs, and for seasons, and for days, and years" (Genesis, 1:14). The Roman poet Ovid's *Metamorphoses,* a compendium of ancient myths, describes a formless chaos, out of which a god brought cosmic order.

Although in our time the study of the stars has been divided into the separate branches of astrology and astronomy, the study of celestial phenomena and the interpretation of their prophetic significance, writes astronomer N. M. Swerdlow in *Ancient Astronomy and Celestial Divination,* "were closely related sciences carried out by the same scholars." Ancient ideas on the nature of life, writes archaeologist Franz Cumont in *Astrology and Religion Among the Greeks and Romans,* were characterized by the fact that they "closely connect belief in the gods with observation of the sky." Thus the earliest scientists were priests, as well, who believed that divinity was revealed in the language of the cosmos. The sum total of this ancient tradition represents a collective book of wisdom drawn from the stars—a form of heavenly guidance to direct the lives of those below.

Indeed astrology, as the *Oxford English Dictionary* defines it, is "the art of judging . . . the occult influences of the stars upon human affairs." This simple definition sums up the soul of a classic tradition that has endured for thousands of years. For in its essence, astrology joins the exoteric science that charts the planetary motions across the sky with the esoteric art of inferring specific meaning from those movements and applying them to human affairs. So persistent has astrology proved to be throughout the course of history that in a lecture he presented during a February 1997 astrology conference, Jungian scholar James Hillman described it as an "archetypal phenomenon" that is ". . . widespread, timeless, and emotionally compelling . . ." Thus if it is archetypal, maintained Hillman, "astrology is here to stay; because it won't go away, it must be archetypal." One reason for astrology's perennial appeal may be its twin roots in both calculated astronomical observation and the art of hermeneutics, or interpretation. Even though the Copernican revolution shifted the ancients' center of the cosmos away from the Earth to the Sun, it did not alter astrology's fundamental premise of reading symbolic import and meaningful design in the visible patterns created by the motions of the planets and stars.

While astrology has metamorphosed over the centuries, adapting itself to the philosophical and religious worldviews of various cultures and historical eras, the story of how it has evolved reveals a surprisingly consistent philosophical structure. The system inherited by Western astrologers, and the one that most people are familiar with today, had its rudimentary origins about four thousand years ago during the Mesopotamian Kingdom of Babylonia. Anxious to ensure that their actions were in accord with divine intention, kings and princes turned for advice to the *ummanu,* a caste of scholar-scientists who examined patterns in nature for glimpses of the future. Ancient "omen texts" preserved from Ninevah's royal library reveal that these priests searched for clues in the flight paths of birds, the behavior of animals, dream interpretation, and especially the stellar events that lit up the night skies. Portents were read in solar and lunar eclipses and other meteorological phenomena. These early "protoastrologers," writes scholar Peter Whitfield in *Astrology: A History,* believed that the sky and the natural world were the pages upon which the gods wrote their decrees in a mysterious language. "The signs on earth, just as those in heaven, give us signals," he writes, quoting an ancient Babylonian scholar. In his book, he cites a cuneiform text from the Old Babylonian period (ca. 1700 B.C.E.) that reveals how the sky reflected the future to those on earth below:

> If the face of the sky is dark, the year will be bad.
> If the face of the sky is bright when the New Moon appears, and it is
> greeted with joy, the year will be good.
> If the north wind blows across the face of the sky before the New Moon,
> the corn will grow abundantly.
> If on the day of the crescent the Moon-god does not disappear quickly
> enough from the sky, disease will come upon the land.

Over time an astral religion began to take shape, and a theology of the stars emerged. The five planets visible to the naked eye, writes N. M. Swerdlow, functioned as "interpreters" whose risings and settings revealed divine intention. The early omen-culture laid the foundation for a more sophisticated form of astronomical calendar-making. Standing on ziggurats—steeply stepped temple towers that functioned as observatories—Babylonian diviners carefully noted the stars' progress across the sky. Driven to decipher the will of the gods above them, they meticulously recorded the movements of the planets and stars in precise monthly "diaries," while noting corre-

sponding events on earth. Over the centuries this priesthood of astrologers compiled tables that allowed them to accurately predict
both the phases and eclipses of the moon—a skill that endowed them
with supernatural powers in the eyes of ordinary people. The positions of the stars, the identification of gods and goddesses with the
planets, and the 360-degree zodiacal circle divided into twelve constellations ruled by animals such as the bull and the crab (the origin
of our popular zodiacal "signs") are among the Babylonian contribu-
tions to astrology.

After the Persians conquered Mesopotamia, the astronomical
techniques and beliefs of Babylonian astral religion passed to Egypt,
India, and Persia, where astrology branched into separate indigenous
traditions. But it was in Hellenistic Greece, particularly the city of
Alexandria in Egypt, that astrology emerged as the tradition that it
most closely resembles today. Scholar Robert Schmidt, a translator of
Greek and Latin astrological texts, writes in his essay "The Problem
of Astrology," on projecthindsight.com, that the astrology developing in Egypt was translated into Greek around 200 B.C.E., resulting in
"a tremendous flowering of astrology during the Hellenistic era that
lasted until the 6th century C.E."

The conditions that led to the rise of Hellenistic astrology began
sometime in 500 B.C.E. Over the next several centuries, wars cat-
alyzed by the Persian conquest of Babylonia and the Alexandrian
conquests resulted in a rich cross-fertilization of ideas between east
and west. The mystery cults of Zoroastrianism and Mithraism, the
Babylonian pantheon of gods and goddesses, and the emergence of
Hellenistic philosophy and science together, writes Peter Whitfield,
created a "complex merging and re-shaping of religion, philosophy
and science."

For astrology, the result of this confluence of ideas was a more as-
tronomically accurate and theologically sophisticated system than
had existed before. The advancements of Greek science, for example, made possible more precise forms of celestial observation. The
concept developed of the cosmos as a series of interlocking spheres,

with the Earth at the center. While fundamentally flawed, this image of the cosmos made it possible to chart the heavens in geometric terms with poles and tropics, an equator and ecliptic. Mathematical relationships between the varying paths of the planets, writes Whitfield, could be charted as they moved along regular geometric paths. The positions of the five visible planets—both in the far past and the distant future—could be predicted with startling accuracy. As a result, a more scientifically based astrology began to flourish. Most contemporary Western astrologers, for instance, base their craft primarily on mathematical techniques taken from Ptolemy's *Tetrabiblos.* This legendary scientist and astrologer, who lived in Alexandria during the second century C.E., produced manuals that sought to explain astrology as a natural science by exploring the physical laws behind the effects of the stars and planets on the Earth.

While Greek science gave geometrical structure and scientific substance to astrology, Greek philosophers discovered in the Babylonian astrological tradition a template for expanding their religious conception of the world. Deeply influenced by the astral religion of ancient Mesopotamia, and lacking a cosmology of their own, the Greeks merged the Babylonian star deities with their Olympian pantheon of gods and goddesses. More important, they adopted the belief that the planets were an embodiment of each of these heavenly divinities, as well as the idea that the eternal cosmos above influenced events on Earth below. At the same time, Eastern beliefs in a transcendent higher power from which individual souls descended and to which they would one day return flowed from the ancient wells of Persian Zoroastrianism, watering the minds of Greek thinkers like Pythagoras and Heraclitus. This fusion of mystical Eastern thought with Greek rationalism flowered in the philosophical works of Plato. The vision of the cosmos sparked by Socrates and fully elaborated by Plato was to inspire astrologers across the centuries.

Plato's cosmological dialogue the *Timaeus,* writes Tamsyn Barton in *Ancient Astrology,* was influential among astrologers in establishing the idea of a relationship between human souls and the stars. In it, Plato envisioned an overarching cosmos that was mathematically ordered and imbued with meaning. Created by a *demiurge,* or divine craftsman, the universe as imagined by Plato was the physical body of a single intelligent being, the *anima mundi,* or world soul. Thus to Plato, all creation was sacred in origin. The stars and planets were divine beings; even the souls of humans had originally been stars, and would one day return to their home in the vaulted heavens where existence was eternal and constant. Plato's philosophy of an intermediary realm of eternal images—archetypes or ideas operating behind the world of appearances—intersected neatly with astrological notions of the planets as divinities.

The mounting preoccupation with fate and destiny in the early centuries before Christ proved another important factor in the evolving role of astrology in Hellenistic culture. Astrology, with its claim to decipher the handwriting of the gods in the sky and its uncanny ability to precisely predict the date of an eclipse, had always offered an alternative to the cruel caprice of blind chance. Those who professed a belief in the astrological doctrine of destiny, writes Franz Cumont in *Astrology and Religion Among the Greeks and Romans,* "elevate to a duty complete resignation to omnipotent fate, cheerful acceptance of the inevitable." This earnest trust in divine fate spread over the Hellenic world. It was enthusiastically embraced by the Stoic school of philosophy, a Greek school of thought that enjoyed popularity among such prominent Romans as Cicero and Marcus Aurelius. According to Stoic teachings, resisting destiny was a source of pain and suffering, while accepting fate brought peace of mind. In addition, the philosophy of determinism made the chaotic events of history easier to bear, recasting them, writes Peter Whitfield, as an "ordered and purposeful flow."

In this way it could be said that astrology infused life, with all its unpredictable uncertainties, with a reasonable and meaningful coherence. As historian Richard Tarnas writes in *The Passion of the Western Mind,* the emergence of the astrological perspective over the course of the Hellenistic era led to the belief "that human life was ruled not by capricious chance, but by an ordered and humanly knowable destiny defined by the celestial deities according to the movements of the planets. Through such knowledge it was thought that man could understand his fate and act with a new sense of cosmic security." Radiating outward from the cultural center of Alexandria, writes Tarnas, astrology became the one belief system that cut across the boundaries of science, philosophy, and religion, embraced alike by "Stoic, Platonic, and Aristotelian philosophers, by mathematical astronomers and medical physicians, by Hermetic esotericists and members of the various mystery religions." As Demetra George notes in her August 2003 article, "The Golden Thread: The Cultural Transmission of Astrology," for the *Mountain Astrologer,* the corpus of Hellenistic astrology generated during this golden era contained within it "all of the fundamental principles and techniques of the Medieval Arabic and Latin, Classical, and Modern traditions." Some of the historical forefathers of astrology at that time, she writes, were Dorotheus of Sidon, Vettius Valens, Claudius Ptolemy, and Firmicus Maternus.

But this intellectually expansive, metaphysical approach to astrology was not to last. As the balance of power shifted in the ancient world from Athens and Alexandria to Rome so, too, did astrology. The geographical shift brought along with it a shift in emphasis, as well. Although there existed learned astrologers who carried on the tradition they had inherited from Greece, astrology began to drift away from its foundation in natural science and philosophy toward power and politics. Roman generals and emperors—such as the emperor Augustus, who, as the first emperor to use astrology to legitimate his position, had his birth sign Capricorn stamped on a coin and distributed throughout the realm—sought to confirm their dominion through manipulating the will of the gods through the stars.

With the decay of the Senatorial Roman Republic and the rise of the Roman Empire, power became increasingly concentrated in the person of the emperor. Swept up in the political intrigues behind the throne, court astrologers cast charts to determine the most auspicious moment for coronations and closely observed the birth charts of those born to positions of potential power. In this highly charged atmosphere surrounding the possession of supreme political authority, astrological "secret police" were on the lookout for potential rivals to the throne. Indeed, writes Tamsyn Barton in *Ancient Astrology,* to be an astrologer in Roman times was a dangerous occupation and could be linked to treason; astrologers predicting imperial futures to rivals of the throne could be executed or banished.

After the triumph of Christianity in the fourth and fifth centuries C.E., pagan practices of astrology, magic, and divination became punishable crimes. Christians were forbidden to pray to the Sun, Moon, or stars. Yet, though the emperor Constantine ordered in the Theodosian Code that "the inquisitiveness of all men for divination shall cease forever," astrology continued to persist throughout the Middle Ages as a powerful underground force. For one thing, references to the stars were written into the Old and New Testaments: in Genesis, God refers to the lights placed in the firmaments as "signs," while the Star of Bethlehem, predicted by the Magi, a term describing Chaldean astrologers, announced the birth of Jesus.

The struggle over astrology in the Christian tradition had less to do with whether it was true or false, however, and more to do with the conflict between the astrologers' belief in fate and destiny and the Church's emphasis on the free will of believers. According to Christian thought, believers were not condemned to live out a preordained fate, but could be redeemed from their sins through repentance and divine grace. The heart of this theological debate centered on the issue of who set the boundaries of destiny. A kind of contest developed that pitted God, or Divine Providence, against the stars. In the eyes of the Church Fathers, for example, to be able to predict the future as it was written in the heavens led inevitably to the idea that the future was fixed—a belief that to the Church Fathers seemed to deny the power of God over earthly existence. Further, belief in an irrevocable Destiny undermined the purpose of worship or prayer.

Even within the tradition of astrology itself, there had been disagreement over whether the stars were the causes of events, or were signs that foretold them. Origen, a native of Alexandria in the third century who spoke Greek, attempted to settle this debate. Though strongly against the practice of astrology, he wrote beautifully of the stars as spiritual beings created by God as signs and intermediaries of His will—rather than causes in themselves. Yet as the Church sought to consolidate its power, writes Tamsyn Barton, it grew less tolerant of rivals; critics such as St. Augustine brought "a new harshness" to the pursuit of heretics like astrologers. As the West slipped into the dark Middle Ages, science, astronomy, and philosophy became eclipsed by the singular authority of the Church. Without the textbooks and treatises of classical Greece and Rome translated into Latin, the intellectual foundation of astrology began to weaken and, eventually, crumble into disuse.

Like a sun dawning over a new horizon, however, the discipline of astrology migrated east. Hellenistic texts made their way to India by the second century C.E.; combined with the wisdom of India's spiritual traditions, the Vedic system took root. Since there was no scientific revolution in India, Vedic astrology—which is based on the "sidereal" zodiac that locates the constellations in slightly different positions in the sky than does the "tropical" zodiac of the West—has retained its consistent form for nearly two millennia. Around the eighth and ninth centuries, as part of a larger transmission of the scientific and literary works of classical Greece, astrological texts were translated into Arabic. Inspired by the knowledge of the Persian astrologers, whose dynasty they had conquered, as well as by how astrology served to link humankind to its eternal cosmic nature, Islamic scholars enthusiastically embraced the practice of astrology. As Demetra George notes in her article, the young Persian astrologer, Masha'Allah, was employed to cast the foundation chart for the new capital of Baghdad. At the legendary House of Wisdom in Baghdad, where the Greek texts of the ancient world were translated into Arabic, scholars pondered the deeper philosophical implications of astrological doctrines. Islamic astrologers further refined the astrolabe—a two-dimensional model of the heavens used to calculate the positions of the Sun and stars. The ninth-century astrologer and philosopher Abu-Mashar, while a teacher at the House of Wisdom, initiated the tradition of mundane astrology—the effect of planetary cycles on world history.

Preserved by Islamic scholars during the bleak centuries of Europe's Dark Ages, astrology reemerged once again in the West during the twelfth century. As the first universities opened across Europe and interest in classical learning spread, Latin translations of Arab translations of Greek astrology texts sparked a revival. Origen's earlier view of the stars as agents of the divine gained greater acceptance among Christians; this time around, recounts Peter Whitfield, the stars were seen as "natural mechanisms devised by God to govern his creation." With paganism a safe thousand years in the distant past, and the authority of the Church firmly consolidated, astrology was no longer considered the threat it had been. Indeed, during the twelfth and thirteenth centuries, writes Barbara Walker in *The Woman's Encyclopedia of Myths and Secrets,* the church "took astrology to its bosom." Pope Julius II chose his coronation date according to astrological calculations, Pope Leo X founded a chair of astrology in a university, and even some cathedrals were decorated with astrological symbols.

Although the debate around free will versus destiny continued, writes Whitfield, astrology could now be "safely treated" as part of a classical tradition that included the philosophy of Aristotle and the poetry of Virgil. In the thirteenth century, the medieval scientist Roger Bacon took up the occult study of the stars as predictors of history. Also in the thirteenth century, no less a figure than the theologian St. Thomas Aquinas seriously pondered the effects of the stars on human behavior, asking in his *Summa Theologica* "whether the heavenly bodies are the cause of human acts?" Accepting that the stars indeed had the power to shape human nature, Aquinas nonetheless believed that these influences could be resisted through free will, writing that "the wise man is master of the stars, inasmuch as he is master of his passions"—a phrase, writes Whitfield, that became a key formula in the long-standing debate around astrology and fatalism.

But it was in Renaissance Italy that astrology came vividly alive, resolving the conflict between free will and destiny in a newly imaginative way. In fifteenth-century Florence, a vibrant artistic and intellectual community abounded. Nurtured by the patronage of the wealthy Medici family, there was a resurgence of interest in Greek and Roman classics. In this milieu, the renowned Renaissance philosopher Marsilio Ficino—priest, physician, and scholar—began translating the works of Plato and other Neoplatonists into Latin. Moreover, Greek texts of astrological writings resurfaced. Steeped in his translations of these esoteric teachings, Ficino began to see astrology as part of a perennial philosophy that had existed since ancient times.

One of the first philosophers to help clients improve their lives through astrology and other occult means, Ficino was among the principal figures responsible for a new direction within astrology. Inspired by the pantheon of gods, goddesses, planets, stars, and the Platonic notion of the world soul, or *anima mundi,* Ficino elaborated a doctrine of natural magic. Departing from prevailing astrological doctrines of fate as fixed by the stars, Ficino envisioned a way in which the sympathies that existed between the cosmos and humans could be positively channeled through talismans, music, and imagery. In other words, rather than use astrology as a technique for prediction, Ficino saw astrology as a powerful form of magic. By drawing down the energies of the stars, astrology could actually be used to *change* the direction of fate—astrology as a magic wand that could alter the pattern of one's destiny, if you will, rather than a crystal ball that foretold a predetermined outcome.

At once practical and magical, this version of astrology strived to give more power to the individual over the blind forces of destiny. Unnerved by the horrors of the Black Death in the fourteenth century, individuals longed for foresight and control over the vagaries of nature. Knowledge of the future provided by astrologers, for instance, could help one prepare an individual for forces he might be facing in the future. In *Cardano's Cosmos: The Worlds and Work of a Renaissance Astrologer*, a study of the life of the Renaissance astrologer Girolamo Cardano, Anthony Grafton writes that Cardano and his fellow astrologers "saw themselves as offering their readers a sharp-edged set of tools with which to free themselves from slavery to external circumstances . . . They felt certain that knowledge of the future would enhance, not diminish, the mastery of the self . . ." So pervasively did astrology determine codes of conduct in Italian society during the Renaissance, writes Grafton, that the Florentine government gave its generals their batons of command at astrologically sanctioned moments, while the prince Leonello d'Este wore clothes of a certain color to draw down favorable planetary influences. Astrology was taught alongside science and philosophy at universities, physicians offered astrological consultations to their patients, and the ancient art of the stars enjoyed widespread acceptance both on the streets and in the most fashionable courts.

Almost as quickly as astrology enjoyed a creative rebirth, however, it began a precipitous and dangerous decline from grace. The Renaissance rediscovery of paganism and the occult mysteries, including astrology, writes Peter Whitfield, offered an older and more exciting vision that began to threaten the authority of the Church. With the onset of the Inquisition in the sixteenth century, the old conflict between human free will and the forces of fate arose once again; only God, said Church authorities, could know the future, and horoscopes were banned. Astrology became branded as heresy, and underwent a repression in Italy. Once taught in universities and considered a respectable practice of scholars and scientists, astrology fell into the shadow of mainstream society. Aside from exceptions such as the seventeenth-century English practitioner William Lilly, author of the classic treatise *Christian Astrology,* the practice of astrology transitioned over the next several centuries from being a religious and royal art of philosophers and kings to a backstreet commercial enterprise aimed at ordinary people.

It was not the Church, however, but the onset of the scientific revolution in the sixteenth century that evaporated astrology's last vestiges of authority. With the discoveries of the forces of gravity, the rotation of the earth around the sun, and the cosmos as infinite, rather than a finite series of spheres, many of the astronomical theories upon which ancient astrology had been based for millennia crumbled. From the scientific perspective, the constellations were nothing more than patterns that appeared in the sky from the perspective of the Earth. The sight of the planets through the telescope revealed mere objects of dust and rock—not divine gods. And in an infinite universe with no apparent beginning and ending, no top or bottom, what real meaning lay in the astrological axiom "as above, so below"? As the lure of the new frontier of deep space beckoned new generations of scientists, the breach between astrology and astronomy widened, eventually severing its ties altogether.

ASTROLOGY REBORN

In his biography of Renaissance astrologer Girolamo Cardano, *Cardano's Cosmos,* Anthony Grafton writes that the historian of astrology confronts "a tradition that lasted many centuries, one that combined remarkable flexibility in application with a durable commitment to a recognizably uniform set of ideas and techniques . . ." Astrology's inherent adaptability allowed it to not only survive its near-demise brought about by the rise of science, but thrive, emerging anew in the last half century of the second millennium in what some might call a second renaissance. Recent decades have witnessed the reconfiguration of astrology, integrating psychology, mythology, and the new sciences, while still retaining its essential, archetypal core: finding myth and meaning in the movements and patterns of the stars and planets.

During the nineteenth century, the mystical doctrines of Madame Helena Blavatsky, the founder of Theosophy, imbued astrology with esoteric significance. The destiny revealed in an individual horoscope was seen as a reflection of the Eastern concept of karma—the belief that the unique shape of one's present life is determined by past actions. Indeed, the modern era witnessed a series of new directions in the field. Fresh methods of interpreting astrology—such as locating psychological archetypes within the zodiac—strengthened it as a belief system. Astrology came to be seen as providing spiritual and psychological insights into human nature, helping individuals connect to their deeper life purpose. In the twentieth century Carl Jung, the Swiss psychologist, provided astrology with an intellectual framework suitable for a world on the cusp of a new age of spirituality.

Jung's language of archetypes—universal patterns that shape the human psyche—was based on the myths of the ancient world that had provided the foundation of classical astrology. An avid student of astrology himself, Jung rediscovered these archetypal myths in the basement of the unconscious, where he believed they continued to exert a powerful influence on behavior. Though the Babylonian, Greek, and Roman star deities had toppled from the visible sky above, they could now be found in the cosmos within. In this inner world, the planets functioned as personifications of the ancient gods and goddesses. Psychologically, they could be read as the various unconscious conflicts, feelings, and tendencies swirling within one's psyche. Mars in conflict with Venus in a person's chart, for instance, equaled the instinct toward aggression versus the desire to be loved; Mercury in harmonious relationship to Jupiter facilitated an open and expansive mind.

From having been a tool to predict the future or interpret or change the will of the gods, the astrology chart now came to represent a virtual theater of the psyche. An entirely new branch of humanistic astrology based on the psychological development of character and individuation arose, initiated by twentieth- and twenty-first-century astrologers such as Dane Rudhyar, Marc Edmund Jones, and Liz Greene. In this system, the astrology chart could be read to reveal a person's innate characteristics and tendencies.

For the professional astrologer, the chart offered a blueprint that could gauge their clients' talents, shortcomings, strengths, and idiosyncrasies.

In recent decades, emerging new paradigms in science that embrace a more holistic and ecological view of nature have inspired some astrologers to take up the quest once again for how astrology might "work" in the physical world. Based on extensive research, astronomer Percy Seymour, Ph.D., for instance, has documented the way the movements of the planets affect the magnetism of the Sun. In his February 2002 article for the *Mountain Astrologer,* "Astrologers by Nature," he writes that "the Sun is transmitting information about planetary motions to the Earth via the solar wind." In addition, Seymour theorizes that over millions of years, the process of evolution instilled in humans "internal clocks that match changes in our light environment and in the magnetic field of the Earth."

Physicist-astrologer William Keepin, Ph.D., has shown how theoretical physicist David Bohm's holomovement theory is relevant to modern astrology. According to Bohm's theory, explains Keepin in an interview with the *Mountain Astrologer* in December 1995, "the flow of reality has two aspects to it—the implicate and explicate order." The explicate world, he says, is the universe of material space, time, matter, and energy familiar to us all. The implicate order, however, is a wavelike information field that interpenetrates each and every point in space-time. Much like Plato's vision of a realm of idea-images, the implicate order, says Keepin, is "a vast realm of meaning and purpose and all of the invisibles and intangibles . . ." Thus astrology, says Keepin, "bridges the implicate and explicate orders more clearly than any other esoteric science that I know of. It ties the whole unfolding evolution of meaning directly with the physical unfolding of the cosmic processes in the form of planets . . ."

Indeed, in searching for a system that unites the multilayered nature of existence, it is perhaps astrology, more than any other discipline, says Richard Tarnas in an interview in the same issue of the *Mountain Astrologer,* that "discloses a universe that is saturated with meaning, with soul, and with an intelligence that is so powerful it is orchestrating synchronicities between the patterns of the heavens and the patterns of human experience . . ."

ASTROLOGY: A MODERN MYTHOLOGY

In his book *Consilience: The Unity of Knowledge,* scientist Edmund O. Wilson writes that nothing in science—or life—makes sense without theory. It is our nature, he writes, "to put all knowledge into context in order to tell a story and to re-create the world by this means." And while scientists test their hypotheses in the "acid washes" of skepticism, tests, and experiments, their theories, he writes, are inevitably a product of *informed imagination.*

Wilson is no fan of astrology. But his observations of how science works could be applied to how astrology works in our time, as well: grounded in the observations of the physical movements of the planets through space, yet imbued with meaning by the imagination of the observer. Thus to the astrologer schooled in both astronomy and mythology, the patterns in the stars reveal narratives and themes that echo throughout the long arc of human history. In the whorls of a seashell, in the cycles of history, and in the human psyche, writes Jungian analyst Michael Conforti in *Field, Form, and Fate,* the recurrence of patterns is a universal presence that exists as an inborn, archetypal blueprint.

Astrology is perhaps unparalleled as a lens for perceiving these archetypes underlying life and nature. For whether looking outward into the infinite darkness of space studded with exploding stars, black holes, and galaxies, or inward to our own complex, chaotic realities, there exists a consistency of pattern and design, story and narrative woven throughout all nature. The orbits of planets, the mathematical alignments between the Earth, the Sun, and each of the planets reveal elegant patterns that simultaneously echo the parameters of myth as well. For though the constellations may exist only as a faint afterglow of the ancient psyche, the "two-million year-old man" within each of us, they are as essential a part of our Earth mythology as the world's major religious traditions. The traditional figures that we represent on our charts, writes the archaeologist Franz Cumont, "are the fossil remains of a luxuriant mythological vegetation."

Thus astrology today could be seen as an archive of ancient wisdom whose roots lie deep in a golden past when the natural world was inseparable from the spiritual. One of the most significant developments in the field of astrology today is the restoration of large portions of its lost history. Beginning in the early nineties, a movement began to translate manuscripts of ancient astrological material from Latin, Greek, and Arabic into English. Thus the works of those great astrologers that centuries ago laid the foundation for modern astrology are now being distributed and studied—some for the very first time. As the astrological tradition reclaims its past and hones its academic credentials by expanding into colleges and universities, and as astrologers become philosophers, scholars, and culture critics in addition to interpreting individual charts, it hovers on the edge of an exciting new era.

Indeed, astrology's basis in both the physical realms and invisible dimensions of the spirit makes it an invaluable aid in repairing the broken thread between past and present, and between the soul of humanity and the soul of nature. Like a symbolic compass, it can be used to help us locate our place in the universe. Observing the transits of the planets through the different signs and houses of the zodiac along the horizon, calculating the aspects and orbs between these planets, noting the cycles of the Sun and Moon weave the thread of individual life into the cosmic tapestry. Working with the planets this way, they are more than spheres of dust and rock. They are characters in a mythic drama of which we, too, are a part.

3.

HE WHO LIGHTS
THE WAY FOR MORTALS:
THE CLASSICAL MYTHS
OF MERCURY

Who can glimpse a god who wants to be
invisible gliding here and there?
—*The Odyssey,* translation by Robert Fagles

Hermes, for to you beyond all other gods it is
dearest to be man's companion . . .
—*The Iliad,* translation by Richmond Lattimore

To the ancients, all of life was a divine play. Every aspect of human experience corresponded to a god or goddess who represented a living image of reality. Venus was the perfume of love and beauty. Mars was the spirit of raw force and aggression. Saturn encoded the limits of time and necessity, while Jupiter radiated power. These colorful characters were endlessly entangled in dramatic conflict with each other, either warring, making love, dancing, feasting, or raising children—just like their human counterparts.

In existence thousands of years across time and culture, the sacred stories of the gods and goddesses reflect the leitmotifs—betrayal, success, tragedy, creativity, death, and birth—that narrate the human condition. Unlike the world's religious teachings, they do not speak of what we *should* be, but what we *are.* Thus the gods, writes the classicist Walter F. Otto in *The Homeric Gods,* represent "the faculty of seeing the world in the light of the divine, not a world yearned for, aspired to, or mystically present in rare ecstatic experiences, but the world into which we were born, part of which we are, interwoven with it through our senses and, through our minds . . ."

Invisible to the naked eye yet powerfully present, these old stories with their familiar protagonists define the longitude and latitude of the psyche's interior geography. An invaluable guide, myths map the dark and light of human nature, the deep background behind our ordinary "tempest-in-a-teacup" psychodramas. For though we may think we have left these phantasmal gods far behind, they live on in the fierce passions and moral dilemmas that shadow our everyday lives. "We are still as much possessed by autonomous psychic contents," wrote Carl Jung in his "Commentary on *The Secret of the Golden Flower,*" as if "they were Olympians."

While modern-day psychology has made great use of classical myths, bringing the psyche to life by naming complexes and neuroses after gods and goddesses, astrology, too, is grounded in the language of mythology. Astrological terms link us to this ancient lineage of images and stories, especially the stars and planets, who were named after the divinities they were thought to be ensouled by. Indeed our planetary names, writes Franz Cumont in *Astrology and Religion Among the Greeks and Romans,* are "an English takeover of a Latin translation of a Greek rendition of a Babylonian nomenclature." The Babylonian gods Marduk, Ninib, Nergal, Ishtar, and Nabu became, respectively, the Greek Zeus, Kronos, Ares, Aphrodite, and Hermes. The Romans renamed these same deities Jupiter, Saturn, Mars, Venus, and Mercury—the original five planets of the astrological tradition we are familiar with today.

Over time, embellished by myth, literature, legend, fairy tales, art, music, architecture, sculpture, and poetry, each planetary divinity accrued layer upon layer of meaning. Like a kind of virtual museum, astrologers today are able to draw upon this collective treasury, using the set of symbolic meanings that has grown up around each planet to amplify the deeper "soul stories" of their clients' lives. The planets and their positions in a natal birth chart can be used to illustrate the particular mythic theme a person may be living, pointing out his or her guiding "daimon," or unique destiny. This principle is colorfully depicted in a drawing in a medieval manuscript. In it, a group of "great men" such as Seneca and Theodosius are shown standing beneath a heavenly sphere; a line links each man to his own special guiding star above.

Accepting our chart for what it reveals we really are, rather than what we think we should be, is one way of creatively working within the confines of fate. It brings imagination to those places where we are most stuck and embittered. Faced with the god whose temple is our chart, we are humbled by the greater forces we have been called to serve. A person with Pluto, the planet of deep transformation, prominent in his chart, for example, might find himself drawn to penetrate the deeper layers of the unconscious psyche, or to pursue positions of power. One woman I know whose chart is strongly influenced by Venus has spent most of her life contending with fateful, compelling relationships; for this woman, love is the central theme of her life. Yet another man whose chart I studied was born with Uranus, the planet of rebellion and unconventionality, in a prominent position: A lifelong political activist, his destiny is to challenge society's status quo.

Each natal chart, however, is much more than a reflection of an individual person. It is the stellar imprint of a moment in time, a page or a chapter in a collective myth in which that person plays a small role in a larger, ongoing drama. Tracking the transits of the different planets across the sky, astrologers take note of those archetypal energies that are at play in the cultural zeitgeist: Is grim taskmaster Saturn opposing powerful Pluto, for instance, as it was during the September eleventh tragedy, stirring tension and strife? Is Venus conjuncting Jupiter, casting a sweet spell over lovers and co-workers? Or is Neptune, maker of dreams and mystic vision, softening Mars's defensive armor?

Like colors or chords, each of the planets strikes a particular note in everyday life. This book, however, is about the cosmic influence of one planet in particular: Mercury, the planet that rules the realm of communications and the way each person thinks, talks, and exchanges ideas with the world around her. By understanding the myths of Mercury, we gain insight into the way Mercury works astro-logically—and the mysteries of Mercury retrograde.

TRICKSTER AND TRAVELING COMPANION

Mercury is the Roman name given to a god whose origins date back to Babylonia, where he was known as Nebo, the god of writing. In ancient Egypt, he was called Thoth, god of secret wisdom, scribe of the gods, and inventor of writing. Mercury's Celtic equivalent was Lugh, the patron of merchants and travelers; he shares similarities as well with the Norse trickster god, Loki.

Before the Roman Mercury, however, there was the Greek god Hermes, the "friendliest of the gods to men" according to classicist Walter F. Otto. Through the medium of Greek myth and classical lit-erature Hermes emerges as a provocative, complex character who played a widely contrasting range of roles. A spell-casting god of magic who could appear and disappear at will, Hermes has fascinated and beguiled scholars, alchemists, astrologers, and poets over the cen-turies. Poetically gifted, the silver-tongued "god of communications" was a figure of puzzling ambiguity who could be by turn charming and deceitful, duplicitous and knowledgeable, clever and sincere. Part mis-chievous devil and part guardian angel, Hermes was also pragmatic patron of merchants and thieves, mystical guide of souls, sympathetic healer, lighthearted companion of adventurers, bringer of sudden luck, and, finally and most important, trusted "messenger of the gods."

In searching for a single strand that weaves the wildly clashing colors of this unusual god into a coherent pattern, the first thing to be said of Hermes is that he is no hero. Indeed, the fleet-footed messen-ger god shares none of the brave characteristics that distinguish Apollo, Zeus, or Hercules. He does not challenge the gods (at least not to their faces) nor does he undertake daring deeds (unless or-dered to). Rather, Hermes achieves his ends through subtlety, humor, inventiveness, and sly craftiness—the mythic prototype of the modern-day spin doctor or the diplomatic mediator. Modern-day tricksters range from the clever advertiser to the political spin doctor, or, in its more noble expression, the diplomatic mediator. And though not a hero himself, these special skills serve to make him an invaluable help-meet and companion of heroes: the puckish and impudent James Carville to Bill Clinton, for example, or the honey-tongued advertising executive Charlotte Beers hired by the Pentagon to help sell America abroad. More venerable examples of Hermes' talent for diplomatic genius include the British statesman Benjamin Disraeli, or President Jimmy Carter, who in his role as global mediator has helped to re-solve disputes between third-world dictators and Western leaders.

The subtle innuendo, the artful turn of a phrase, the charming half-truth, and the coy, perfectly timed bon mot: These are the un-cannily human traits of an antique god who still speaks to us across the centuries. Hermes' capricious, many-faceted character is color-fully depicted in the *Homeric Hymns*. Mythically attributed to Homer, and composed between 800 and 300 B.C.E., these archaic psalms to the gods and goddesses from a long-vanished era are classic, timeless pieces of prose precisely because they describe characteristics still alive in human nature. The "Hymn to Hermes" describes the god's divine genealogy and tells stories that illustrate his quicksilver, imp-ish nature. Of illegitimate heritage, Hermes is born from the illicit li-aison between the nymph Maia and Zeus. Maia is one of the "star sisters" in the constellation Pleides whose appearance in the night sky in mid-May signals the movement of the Sun into Gemini—the sign ruled by Mercury.

From his very first breath, recount the Hymns, little Hermes is rest-less and inquisitive. A precocious infant, he is instantly curious to dis-cover the whereabouts of the cattle belonging to his older half brother, Apollo. Jumping from his cradle, Hermes bounds out of his cave, only to immediately encounter a tortoise. Beguiling the turtle with clever promises into returning with him to his cave, the enfant terri-ble scoops out the flesh of the unsuspecting turtle, then handily crafts the first lyre out of oxhide, sheep-gut strings, and the tortoise's shell.

Quickly tiring of his new musical invention, Hermes sets off once again on what will become the first road trip down many a winding path. In one of the most famous parables in the lore surrounding Hermes-Mercury, our mischievous god next comes across Apollo's herd of cattle—and makes the decidedly impudent decision to rustle them. But it is *how* Hermes steals his half brother's cattle that is so revealing of his trickster nature: He leads them away backward, so that their hoofs point in the opposite direction from the cave where he has hidden them. Further, he disguises his own involvement by crafting a pair of sandals that magically cover all trace of his own footsteps. Hermes is not in this for the money, but simply for the thrill of the dare. With little real interest in the cattle, Hermes kindles a fire, roasts two cows, and, in a neat bit of astrological symbolism, divides them into twelve portions as a sacrifice to each of the gods.

Soon after, Hermes returns home and, according to the hymn, transforms himself into an airy mist, enters his house through a keyhole, crawls back into his cradle, lyre under his arm, and falls sound asleep. When Apollo discovers the theft of his cattle, he immediately suspects his new baby brother. Cleverly taking refuge in his infancy, Hermes loudly proclaims his innocence. But Apollo knows otherwise, and, hauling his naughty sibling before his father, demands accountability.

Amused by Hermes' charming protestations, Zeus, perhaps like any mortal father, simply laughs and tells the boys to get over their quarrel and make up. Having successfully manipulated his father, Hermes relents and confesses to his brother the location of the stolen cattle. Turning his irresistible charm on Apollo, Hermes enchants him with a performance on his new lyre. Suitably softened, Apollo strikes a deal with Hermes, whereby his little brother will teach him to play the lyre in exchange for sharing dual patronage of the cattle. In a gesture of reconciliation, Apollo bestows upon his baby brother a staff—Hermes' trademark magic golden caduceus. Entwined with two serpents, the staff symbolizes his divinatory powers, as well as the god's ability to harmonize two conflicting viewpoints. (See Figure 1.)

FIGURE 1

The Caduceus of Hermes

Chambers Encyclopedia—Vol.II, (J. B. Lippencott Company)

The 2,600-year-old hymn highlights Hermes' more amoral characteristics as curious and guiltless child, imaginative deal maker, inventive craftsman, and charming sweet-talker. In it, we see clearly the evolution of the trickster archetype, mischievous "master of secret power, who causes both finding and losing," as the mythographer Karl Kerenyi describes him in *Hermes: Guide of Souls*. The concept of losing then finding something leads directly to another aspect of Hermes—that of the helpful companion who accompanies lonely travelers along the way. (See Figure 2.)

The Greek *herm,* the origin of Hermes' name, means "he of the stone heap." Throughout the ancient Greco-Roman world, short pillars with Hermes' head at the top and penis on the front were set into stone heaps at the crossroads of major thoroughfares. Travelers pass-

A Greco-Roman *herm*

A Dictionary of Classical Antiquities (Swan Sonnesschein & Company, Ltd)

FIGURE 2

ing by added stones to these heaps as offerings petitioning Hermes' blessings on their journeys. Hungry wayfarers who pilfered food found at these wayside shrines considered them windfalls, or *hermaion,* from the god of luck. To take this found sustenance as an accidental gift rather than stolen goods—just as the god himself would have done—was thought to be perfectly in keeping with the spirit of Hermes. The traveler who stumbled upon any syncronicity or stroke of fortune—a coin in the dust, or the helpful stranger who appeared out of nowhere—offered thanks to the god Hermes for his divine intervention at the right moment in time. The sudden and surprising twists of fortune he bestowed upon both the good and bad, rich and poor, powerful and downtrodden forged his reputation as the god who came closest to sharing in the travails of the human condition. For his generous assistance to the lost and lonely, Hermes came to be seen as the god who was the friend of mortals as they traveled the road of life. Like the twentieth-century writer Jack Kerouac—whose restless wanderings with his fellow hitchhikers were immortalized in his classic beat novel, *On the Road*—Hermes invoked the heady exhilaration of life as an open-ended trip down the road less traveled. As patron of the journey with no fixed destination, Hermes embodied the capacity to live life with passionate spontaneity and intense intimacy. "For all to whom life is an adventure—whether an adventure of love or of spirit—he is the common guide," writes Kerenyi.

As the guide of wayfarers whose shrine was placed at the intersection of highways, Hermes gradually came to be seen as the god who ruled all forms of transition—whether physical or psychological. Lord of the road, god of thresholds, doors, boundaries, ways, and entries, Hermes-Mercury is thus patron of those who find themselves caught in the interlude "in-between"—between youth and middle age, jobs or marriages, or life and the afterlife. In this twilit state between what has been and what is to come, the heroic defenses of the ego are softened. Letting down our guard, we open to mystery, becoming receptive to the images, ideas, visions, and dreams that arise like spirits from the deeper dimensions of life.

As surreal as this suspended reality may seem, it is familiar territory to Hermes. In *In Midlife,* for instance, the Jungian psychologist Murray Stein describes Hermes as the "edgeman"—the god who exists outside the fixed and familiar boundaries of the daylight world. Trusted guardian of the borderland realm of dreams, Hermes was the god to whom the ancients poured out their last libation before retiring to sleep, and the god to whom reverence was paid after a significant dream. Night itself, points out Walter F. Otto in *The Homeric Gods,* is a metaphor that captures the essence of Hermes. Only through understanding the night—where danger lurks and a robber or ghost may appear at any moment to threaten one's safety—can we fully understand the realm ruled by the god of thresholds:

> A man who is awake in the open field at night or who wanders over silent paths experiences the world differently than by day. Nighness vanishes, and with it distance; everything is equally far and near, close by us and yet mysteriously remote. Space loses its measures. There are whispers and sounds, and we do not know where or what they are . . . There is no longer a distinction between the lifeless and the living, everything is animate and soulless, vigilant and asleep at once . . . But the darkness of night which so sweetly invites us to slumber also bestows new vigilance and illumination upon the spirit. It makes it more perceptive, more acute, more enterprising. Knowledge flares up, or descends like a shooting star—rare, precious, even magical knowledge. And so night, which can terrify the solitary man and lead him astray, can also be his friend, his helper, his counselor."

GUIDE OF SOULS

From the unique constellation of Hermes as trickster, patron of the journeyer, and spirit of the night, emerges Hermes in one of his most enduring and memorable aspects: guide between worlds, shepherd of souls into the afterworld and realms of the spirit.

In the last book of Homer's classic tale of *The Iliad,* and especially throughout *The Odyssey,* Hermes' tricksterish persona is more muted. Instead, he evolves into the wise messenger god of Olympian legend. He sublimates his gleefuly roguish nature, and the more serious, metaphysical dimension of his character emerges. Hermes' appearance at the tragic conclusion of the Trojan War signals the shift away from the heroic world of *The Iliad* toward the great journey epic, *The Odyssey*—a world, writes the mythographer Karl Kerenyi, where Hermes is god. From dealing with matters of trade, travel, and trickery, Hermes turns now to the eternal questions of life and death.

Hermes makes his pivotal appearance in the twenty-fourth book at the close of *The Iliad.* Unlike the intense buildup to battle that has marked the preceding chapters, this moment comes in the aftermath of the Trojans' bitter defeat at the hands of the Greeks. "It falls in the trough of the aftermath," writes Murray Stein of this haunting final episode, "in the period of mourning the losses brought about by the war's unpredictable fury. The battles of the epic are now over, as are the heroic games . . . now begins the time of existing *without* the slain heroes . . ." And so the stage is set for the entrance of Hermes, the helpmeet of heroes.

Book twenty-four opens with the Greek hero Achilles in a rage. His best friend, Patroclus, has been murdered by his enemy, the Trojan soldier Hector, during battle—and now Achilles has killed Hector in revenge. Refusing to give Hector a proper burial, Achilles is savagely mutilating his corpse, dragging it behind his chariot around and around Patroclus's grave. Looking upon this scene, the gods—especially Zeus—are understandably appalled. So, too, is Hector's father, the Trojan king Priam, who undertakes a night journey into enemy territory to rescue the body of his favorite son in order to provide him with a proper burial.

Feeling a pang of pity for the old man, Zeus orders Hermes the Wayfinder, he "who lights the way for mortals," according to the 1975 translation by Robert Fitzgerald, to accompany the aged and suffering king. Immediately upon hearing the mission assigned him by the gods, ". . . the Wayfinder obeyed. He bent to tie his beautiful sandals on, ambrosial, golden, that carry him over water and over endless land on a puff of wind, and took the wand with which he charms asleep—or, when he wills, awake—the eyes of men. So, wand in hand, the strong god glittering paced into the air." (See Figure 3.)

Hermes materializes before Priam out of nowhere as he stops to water his horses at a river. Appearing as a young man standing in the road, the sight of the stranger causes the hair to stand up on the king's arms and legs. Sensing divine intervention, Priam entrusts himself to Hermes, who stealthily spirits Priam through the enemy camp. Drawing on his powers as an enchanter, Hermes showers a "mist of slumber" on the sentries. At Achilles' door, Hermes reveals

FIGURE 3

The Classical Mercury

Council of the Gods
(Thomas R. Rockwell Company)

his true identity as a god to Priam, then vanishes: "I am no mortal wagoner, but Hermes, sir . . . Now that is done, I'm off to heaven again." Due to Hermes' wonder-working spells and gifts as a mediator, all goes well in Priam's negotiations with Achilles, who, finally moved to compassion, agrees to deliver Hector's corpse to the grieving father. Yet if Agamemnon, king of the Greeks, should recognize Priam, says Achilles, he will lose his right to bury his son. To prevent this from happening, Hermes reappears once again in the dead of night and, under a spell of invisibility, safely shepherds Priam and the wagon bearing his son's corpse out of the encampment and back to the river—whereupon Hermes disappears once again. *The Iliad* concludes with Hector's burial rites.

Hermes' role as spirit guide and helpful companion figures even more strongly in *The Odyssey.* The god's connection to matters of the dead and the process of death and dying is evocatively portrayed in the last book of this great epic. Book twenty-four, *Peace,* opens on the heels of Odysseus's triumphant but bloody return to Ithaca after his arduous, decade-long journey home. In a horrific battle Odysseus has slain the men, or suitors, who had occupied his estate and wooed his wife in his absence. Now it falls to Hermes to summon forth the lost souls of the suitors, who wander in limbo between life and death. Robert Fagles's 1996 translation of that scene eloquently captures Hermes in his role as a *psychopomp,* or guide to the afterworld:

Now Cyllenian Hermes called away the suitors' ghosts,
holding firm in his hand the wand of pure gold
that enchants the eyes of men whenever Hermes wants
or wakes us up from sleep.
With a wave of this he stirred and led them on
And the ghosts trailed after with high thin cries
As the bats cry in the depths of a dark haunted cavern,
Shrilling, flittering, wild when one drops from the chain—
Slipped from the rock face, while the rest cling tight . . .

So with their high thin cries the ghosts flocked now
And Hermes the Healer led them on, and down the dank
Moldering paths and past the Oceans' streams they went
And past the White Rock and the Sun's Western Gates
    and past
The Land of Dreams, and they soon reached the fields of
Asphodel where the dead, the burnt-out wraiths of mortals,
make their home.

Down, down, down, descend the fluttering souls, led by Hermes into the cavernous realm of the dead that lies concealed beneath the land of the living. Here the suitors are greeted by the shades of the once-great heroes of the Trojan War, Achilles and Agamemnon among them, who are bitterly recollecting their tragic fates on the battlefield. "As they exchanged the stories of their fates," recounts Fagles's translation, "Hermes the guide and giant-killer drew up close to both, leading down the ghosts of the suitors King Odysseus killed."

Here we see Hermes present at the moment of judgment said to occur at the moment of death, linking him to Anubis, the Egyptian god responsible for weighing the hearts of the dead in the afterworld. Yet Hermes is unique in mythology, as he not only leads souls down into the underworld but also leads them back up into the day world again. It is Hermes, for instance, who returns Persephone from her sojourn with Pluto in his kingdom of the dead. In an ancient vase painting, Hermes is shown standing before a large vessel protruding from the earth, staff raised, winged souls fluttering forth. And to Hermes falls the task in the myth of Orpheus and Eurydice of leading Eurydice up from the land of the dead as she follows her beloved back to life— and then back down again, when Orpheus makes the fatal mistake of looking back. Thus whether fetching Demeter for a meeting with Zeus on the peaks of Mt. Olympus, hovering at the crossroads, or gently shepherding souls into the next world, Hermes is truly the "go-between god," linking the heights and depths, the human and divine.

MAGICIAN AND MYSTIC

The cap of invisiblity that enables the messenger of the gods to slip from place to place unseen, the golden wand that gives him the power to awaken or put to sleep, and the winged sandals that lift him high on his journeys across land and sea all confer upon Hermes the role, as Walter F. Otto writes, of "archwizard and patron of magic."

At a critical juncture in *The Odyssey,* for example, Odysseus sets out to rescue his men from the sorceress Circe's house. Under her spell, they have been turned into swine. At a crucial moment, Hermes appears suddenly at Odysseus's side to warn him of the dangerous bewitchment awaiting him. Digging up a magic herb from the soil, Hermes offers it to Odysseus as protection against Circe's enchantment, allowing him to safely rescue his men and be on his way once again.

Granted the gift of prophecy by his older half brother, Apollo, Hermes possesses oracular powers, as well. Taught by the Triae, three archaic bee-goddesses, the young Hermes picked up the art of augury using pebbles and knucklebones; he is also associated with fortune-telling through casting lots and dice. For those who divine the future, it is Hermes who parts the veil between worlds, allowing a glimpse into the past and the future. In his role as guide of souls, Hermes is naturally poised to communicate with the dead, channelling the spirits of the departed to their loved ones. "Hermes, he-god, who can go in and out of the ground, he-god still wary to keep power with the fathers, now I have need of you, back home from long exile. This is the gravemound of Agammenon, my father. Help my cries through the ground to his ghost," cries Orestes in *The Oresteia,* by Aeschylus (Tony Harrison and Rex Collings, trans.) as the grieving son invokes Hermes to help him communicate with the ghost of his dead father.

Channelling the spirits of the dead, divining the future, breaking or casting spells are classical forms of magic. As strange as it may sound to the modern-day person, however, such ordinary activities as writing and speaking were once also considered the province of magicians. In the ancient world magic and knowledge were closely intertwined, as scholarship and intellectual ability were considered extraordinary and even supernatural talents.

"Divination and the written arts are similar in some respects in Hermes' realm;" writes Freda Edis in *The God Between: A Study of the Astrological Mercury,* "his invention of the alphabet and its association with augury in early cultures helped men and gods to become aware of the . . . workings of the human mind." Hence, the flash of a sudden insight, the ability to translate a nebulous intuition into galvanizing words, or the enchantment of a powerfully articulated story are all transformative techniques that come under the dominion of Hermes the magician.

The link between wisdom, knowledge, and magic came together in the mythic figure of Hermes Trismegistus—a legendary magus of esoteric wisdom. His cult arose, writes Edis, during the four centuries straddling the start of the Christian era when "the Mediterranean world was a ferment of doctrines and sects." In the milieu of that time the Greeks developed a fascination for Thoth—the ibis-headed Egyptian god of writing and magic who stretches into the deepest recesses of time. Impressed by Thoth's similarity to Hermes, they named him "thrice great" Hermes, or Trismegistus, which signified that he combined the triple roles of king, philosopher, and priest. Though he was not in fact a real historical figure, a group of texts called the Hermetic treatises appeared that purported to be the teachings of this mythic great teacher to his disciples. Thought to be the actual work of Greek thinkers who were part of an esoteric school in Hellenistic Egypt, the texts took the form of dialogues on topics such as alchemy, magic, astrology, and philosophy. Their aim was the liberation of the soul through esoteric knowledge and magical techniques—especially the practice of astrology.

In these mystic texts we see the development of Hermes as a sage and spiritual figure. In this role Hermes is the teacher who conveys truths from the unseen realms, dispensing the perennial wisdom of the ages to seekers. Centuries later, when long-buried Greek texts reemerged during the Italian Renaissance, the figure of Hermes Trismegistus reappeared, brought to light by the Renaissance scholar Marsilio Ficino. So revered a figure did he become that some argued his wisdom was of more ancient origin than even Moses. The Hermetic teachings inspired the sages of the time as they delved into the mysteries of alchemy—the process by which lead is changed into gold. In the alchemical tradition of the Renaissance, the spirit of *Mercurius* symbolized the very process by which this transformation occurred.

In our time, the transformation of the base material of lead into gold became symbolic of the inner psychological growth that occurs whenever we consciously "process" the unexamined thoughts and shadowy instincts in the underworld of the psyche. As formulated by Carl Jung, it is through the alchemical interaction between the conscious and the unconscious—the give-and-take dialogue between the different sides of ourselves—that wholeness of self is attained.

No higher form of magic exists, in fact, than the transformation of the human personality from its base and limited nature to the greater soul latent within each of us. In this, Hermes is truly our guide. For this ancient god—who contains within his being both petty thief, mischievous prankster, loving friend, and wise spiritual teacher—is *us.* For this reason he embodies the ambiguity that lies at the heart of the human condition. No comfortable black-and-white deity is Hermes, offering clear-cut ethical forms of behavior. Rather, he is a mercurial god whose mythic legacy exemplifies humankind's struggle to come to terms with a nature that is both flawed and divine at once.

This makes of Hermes a uniquely modern deity, well suited to deal with the numerous ethical quagmires we are forced to deal with today. It is to the myths of Hermes, for instance, that we can turn for insight into the often-amoral dealings of seemingly upright business leaders and entrepeneurs, caught in their shadowy, backroom dealings—or the quick turn of profit conducted in ordinary commercial transactions. Seeing the trickster at work in the world this way, we can turn within and find the shadow of our own inner spin master and deal maker. Hermes is the very image of the contradictory forces that tear at each of our souls: both good and bad, dark and light, wise and frivolous, right and wrong. Through confronting the ambivalent, murkier side of ourselves this way, we can also find within ourselves another, better side of our human nature—the divine companion who wisely learns from mistakes, transmuting the messy chaos of human experience into the light of wisdom.

4.

THE ASTROLOGICAL
MERCURY

Mercury ☿ hath significance of all literated Men; as
Philosophers, Astrologers, Mathematicians, Secre-
taries, Schoole-masters, Poetes, Orators, Advocates,
Merchants, Diviners, Sculptors, Attorneys, Accomp-
tants, Sollicitors, Clerks, Stationers, Printers, Secre-
taries, Taylors, Usurers, Carriers, Messengers,
Foot-Men . . .

> —John Gadbury, *Genethaialogi: The Doctrine
> of Nativities, 1658*

Name: He is called Hermes, Stilbon, Cyllenius,
Archas. Mercury is the least of all the Planets, never
distant from the Sun above 27.degrees; by which
reason he is seldom visible to our sight . . .

> —William Lilly, from *Christian Astrology*

In astrology as in myth, Mercury's function as a planet is that of
message bearer and go-between. Just as the fleet-footed herald
translated Zeus's will to mortals, so, too, does the planet Mercury
astrologically symbolize the process of both receiving and communicating thoughts and information.

Its position orbiting between the Earth and the Sun reflects Mercury's symbolic role as that of intermediary. Like Venus, whose orbit
also lies between the Earth and the Sun, Mercury is considered one
of the inner "personal" planets that govern the more intimate, rela-
tional areas of life. In an individual chart, Mercury could be said to
be the spokesperson of the individual ego. Without Mercury, we

could not communicate our thoughts to others, but would be trapped in autistic silence. At the same time, Mercury reconnects us to our deeper core, our inner self. As well as being the guide between the realm of the living and the realm of spirits in myth, he serves the same role in our psyche, facilitating the exchange of contents between the conscious and unconscious. Thus the planet Mercury is said by astrologers to rule the mental and linguistic activities of learning, thinking, analyzing, concentrating, and communicating.

So critical is Mercury's role in the planetary pantheon that some say it is the most significant planet in the chart. Through Mercury, the autonomous "I" extends itself to the surrounding environment. In a never-ending feedback loop, the self becomes aware and conscious of its surroundings, then is altered and transformed by its perceptions. Think of the insatiably curious child, for instance, who takes apart an old clock, delightedly touching and examining each small part—its mind expanding all the while in a kind of kinesthetic learning experience. Or, consider the poet who, moved by the beautiful image of a brilliant sunrise, is able to translate that wordless experience into a vivid piece of prose.

Mercury is known as the "idea planet," and its tracks are laid out everywhere in our daily lives. The ruler of the everyday kinds of "give-and-take," "back-and-forth" small talk, this is the planet that connects the individual to the world at large. Through the medium of Mercury, the scholar or student studies philosophy, the businessperson brokers a sale, the diplomat negotiates a peace treaty, the teacher conveys a lesson on history, the journalist pens a feature article, the radio host interviews a guest, the advertiser crafts an ad campaign, the minister delivers a sermon, the psychologist interprets a client's dream, lovers discuss their relationship, or two people gossip or debate politics over coffee. In addition to the processes of communication, Mercury governs as well the very *stuff* and technology of communication: a fax machine, a car, an airplane, a journal, a cell phone, a book, a newspaper, or a computer. Whether magazine, commuter train, or news broadcast, whatever takes us from one place to another or links two or more things is governed by the planet Mercury.

Astrologers sometimes illustrate the meaning of a planet through its rulership of a specific part of the physical body. Mercury is said to rule the nervous system, the lungs, shoulders, arms, and hands. In the nervous system, we see the function of Mercury as transmitter, sending signals throughout the body from the brain. Mercury, in fact, is often referred to as the link between body and mind. With the help of our lungs, we are able to articulate our thoughts in sound and speech. With our shoulders, arms, and hands we literally reach out to the tactile environment about us—gesturing, typing, stirring food, turning pages, or touching as a form of sensory feedback, learning and communicating through our very fingertips. Mercury has a general rulership over the senses as well, write astrologers Francis B. Sakoian and Louis S. Acker in *The Importance of Mercury in the Horoscope,* as people with "exceptionally good eyesight, touch, feeling, or sense of smell are apt to have a good Mercury."

In the zodiac, the planet Mercury is assigned rulership of two signs, Gemini and Virgo. Both of these signs are stories that add yet more layers to our understanding of the astrological Mercury. Gemini, sign of the twins, for instance, represents Mercury's fluid, quicksilver, and dualistic characteristics. A witty, sparkling conversationalist, Gemini is easily bored unless infused with new ideas, people, and projects. Ever curious, Geminis crave mental stimulation the way others crave money or power, restlessly dancing from idea to idea. "Unpredictable, restless, and changeable; observant, clever, and curious, too; quick-tempered and destabilizing; playful and talented; funny and friendly—if the desert wind were a zodiac sign," writes astrologer Dana Gerhardt in "A Thousand and One Gemini Nights" in the November 1999 issue of the *Mountain Astrologer,* "there's only one it could be." If we meet Gemini's spirit in the natural world through the wind's movement, writes Gerhardt, we "meet Gemini's spirit in the inner world—no matter our horoscopes—via the movement of the mind." Perhaps it is for this reason that the Chinese call Gemini the "monkey" sign.

Gemini's glyph, or astrological symbol, is composed of two short parallel lines joined by lines on the top and bottom— ♊ —a little like the Roman II. To express Gemini's twin nature, astrologers often turn to the Greek mythic figures of Castor and Pollux, known as the Dioscuri, and represented by the two bright stars in the constellation of Gemini. Astrologer Brian Clark relates the narrative in the June 2000 *Mountain Astrologer* article "Gemini: Searching for the Missing Twin." In it, he tells the tale of Leda, queen of Sparta and wife of Tyndaraeus, the king of Sparta, who was ravished by Zeus in the shape of a swan. It was tricksterish Hermes who had given Zeus the idea to trick Leda by assuming an animal form. Simultaneously impregnated by both Zeus and her husband Tyndaraeus, Leda gave birth to two giant eggs. From these were born two sets of twins: Helen and Pollux, divine progeny of Zeus, and Castor and Clytemnestra, the human offspring of Tyndareus. "One set of twins is divine;" writes Clark, "the other set is mortal. One set of twins is male; the other is female. One set of twins is male-female, divine-mortal, as is the other set."

The myth of the twins around which Gemini revolves reveals its androgynous, shape-shifting tendencies, his—or her—capacity to be by turn both masculine and feminine, human and divine. Gemini's "twoness" is reflected in other ways, as well. Notoriously fickle minded, even shallow, Gemini is equally capable of penetrating intellectual depth—possessing a dazzling breadth of style ranging from the *National Enquirer* to the *New Yorker.* And while Gemini can be the capricious boy-god, or *puer,* he is also the *senex,* or wise old man or magician. Some speak of Gemini's double sidedness in terms of the "good angel" and "bad angel" who sits on either shoulder whispering in both ears. But the myth of the Dioscuri expresses an even deeper archetypal truth: the notion that each of us is born "twinned," separated at birth from a "missing half"—our divine sibling. In Oriental mysticism, as in Jungian psychology, wholeness comes from integrating that part of ourselves that we have split off from consciousness, whether a man's inner feminine or a woman's inner masculine, or each person's soul or Higher Self. Much of Gemini's restless nature, in fact, stems from the search for its missing half. In *Astrology, A Cosmic Science,* the esoteric astrologer Isabel Hickey describes Mercury as the link between the soul and the personality, and between the "heaven and earth" within each one of us.

Virgo, the other sign under Mercury's rulership, is more expressive of the feminine, practical side of Mercury, as well as being noted for its skills of analysis and craftsmanship. Whether deconstructing a complex idea or stitching together a quilt, Virgo is proficient at taking things apart down to the tiniest detail, and then reassembling all the parts in a new design. While Gemini rules the third house in the zodiacal wheel—media, early education, siblings, writing, and communications—Virgo rules the sixth house of work and service. Indeed, the Virgoan dimension of Mercury represents the planetary god's role as servant and helpmate to the gods; in the zodiac, Virgo's reputation is modest, trustworthy, and unassuming, yet highly efficient when it comes to accomplishing assigned tasks.

The ancient association of the golden caduceus with Mercury—the serpent-entwined staff that is still used as a symbol of the medical profession—symbolizes the planet's Virgo function as healer. Journeyer to the underworld, Virgo goes in search of the soul medicine that heals the incurable wound or disease. As the connecting link between body and mind, Virgo heals by making whole, by locating and putting back together those pieces that have become broken or split apart in tiny fragments. The symbolism of the caduceus, says astrologer Laurence Hillman, "signifies that to have health we must balance the yin and the yang energies represented by the serpents. Mercury-Hermes, as the bearer of the staff, becomes the unifying element that holds together the opposites."

VIRGO, GEMINI, AND MERCURY
IN THE INDIVIDUAL CHART

Those who are not Geminis or Virgos may think that Mercury has little influence over their lives. Yet the zodiac wheel that is our individual birth chart reveals that all of us have the planet Mercury at work
somewhere in our lives. In addition, Gemini or Virgo governs specific areas of interest, or "houses," such as work, family, or religion.
The planet Mercury, in fact, comes even more sharply into focus
when we study the unique way it is placed in a person's chart or as
we observe its daily transits through the zodiac. As the seventeenth-
century Christian astrologer William Lilly colorfully summed up Mercury's "dignified" and "ill dignified" manifestations in his *Christian
Astrology*:

> Being wel dignified, he represents a man of a subtil and poli-
> tick brain, intellect, and cogitation; an excellent disputant or
> Logician, arguing with learning and discretion, and using
> much eloquence in his speech, a searcher into all kinds of
> Mysteries and Learning, sharp and witty, learning almost any
> thing without a Teacher; ambitious of being exquisite in every
> Science, desirous naturally of travel and seeing foreign parts: a
> man of unwearied fancy, curious in the search for any occult
> knowledge; able by his own Genius to produce wonders;
> given to Divination and the more secret knowledge; if he turn
> Merchant, no man exceeds him in a way of Trade or invention
> of new wayes whereby to attain wealth.
>
> When ill dignified: A troublesome wit, a kinds of Phre-
> netick man, his tongue and Pen against every man, wholly
> bent to spoil his estate and time in prating . . . a great lyar,
> boaster, pratler, busibody, false, a tale-carrier . . . easie of
> beleef, an asse or very ideot, constant in no place or opinion,
> cheating and theeving every where; a news-monger, pretend-

ing all manner of knowledge, but guilty of no true or solid learning; a trifler . . . constant in nothing but idle words and bragging.

Modern astrologers, of course, interpret Mercury's placement in a more nuanced and psychological fashion. Its position in my own chart, for instance, is an example of how this planet can signify a person's career or calling. On the day of my birth, Mercury was in the sign of Scorpio. This gives my mental outlook a mystical or occult overtone, directing my thoughts toward hidden or esoteric topics. Appropriately for a writer, my Mercury is placed in the third house, Mercury's natural home and the area that governs publications, the media, and writing. Thus my personal Mercury plays a role similar to that of Hermes' mythic role as psychopomp, or guide to the underworld. In other words, my vocation writing about spiritual, psychological, and occult topics—translating abstract, difficult-to-grasp ideas into ordinary language—is expressed through the astrological symbols in my chart.

No two Mercuries, however, are alike. One of the most valuable gifts of astrology is how clearly it images the unique spirit of each person. It takes 26,000 years for the astrological pattern that we were born under to repeat itself. Thus there are as many Mercuries, or ways of thinking and communicating, as there are individuals. Current research by educational psychologists emphasizes that not all children learn in the same manner. In this way, the vast array of Mercury placements and possible patterns reveals the dazzling diversity of "frames of mind" or kinds of intelligence. One man I know, for example, has Mercury in the practical and sensual earth sign of Taurus in the tenth house of career. A gifted businessman, this man possesses the seemingly magical skill of summoning money from the depths, enriching his life and his family's with Taurean comforts of food, furniture, security, and clothing. Another young man has Mercury in the inventive and futuristic sign of Aquarius in the ninth house of philosophy and higher education. He is a lifelong student, an enthusiastic scholar who has just entered graduate school to become an environmental scientist. With his prominent Mercury, he lives to study, write, and talk about ideas.

The position and aspects to our natal Mercury can reveal early childhood experiences around learning. One woman who came to me for an astrology reading poignantly related the lifelong aftereffects of her negative experiences as a young schoolgirl. Born into a highly intellectual family, fearful of her harshly critical teachers, she continued to suffer long into adulthood feelings of insecurity and inadequacy about her intelligence. When I looked at her chart, I saw that she had the planet Mercury in the sign of Virgo, conjoined to the highly spiritualized planet Neptune, also in Virgo, in the third house. As I interpreted her chart, I could see that the planet Neptune may have cast a dreamy and imaginative spell over her thinking processes, making it difficult for her to process information in a linear way. Yet although her nebulous Mercury may have been out of place in a rational, logic-based educational environment, she possessed a powerful *spiritual* intelligence. In fact, she is a hands-on healer, widely admired for her gifts of touch and empathy.

MERCURY IN THE SIGNS AND ELEMENTS

The sign Mercury was in on the day you were born can reveal much about your own personal communications style, the emotional tenor of your thoughts, the subjects that interest you, how you learn, and even the kind of work you are drawn toward. Below is a description of Mercury as it appears in each of the twelve zodiac signs (a table is included in the appendix that lists the positions of Mercury from 1910 through 2025). Each of the twelve signs is modified by what astrologers call its "modality." The three modes are **cardinal,** or pioneering and forward moving; **fixed,** or stable and stubborn; and **mutable,** or adaptable and changeable. In addition, each sign falls in one of the four elements: **fire,** or passionate and energetic; **earth,** or fertile and enduring; **air,** or curious and intelligent; and **water,** or empathic and sensitive.

As astrology is one part astronomy and many parts mythic imagination, remember that these interpretations are merely a rudimentary guideline to help you connect to the unique thinker and communicator within yourself. The set of meanings associated with Mercury in each sign can be applied to the day-to-day position of Mercury, as well.

Mercury in Aries As the first sign of the zodiac, Aries is an initiator and an instigator. Aries thinkers are "thought pioneers" who feel most at home exploring uncharted frontiers—either inner or outer. Extroverted and friendly, those born with Mercury in the **cardinal fire** sign of Aries are direct and outspoken in their communications with others. Don't expect a Mercury in Aries to keep his thoughts to himself; as soon as something pops into his mind, he is likely to blurt it out rather than mull it over. A fire sign, Mercury in Aries' tendency to forthrightness can sometimes seem rude, yet is more likely refreshingly honest, with a talent for clearing the air of unspoken and unpleasant thoughts. A dynamic sign, those with Mercury in Aries are enthusiastic, positive, optimistic, and upbeat. Quick thinkers in whatever field they choose, they can be daring and challenging to those with outmoded opinions and thoughts. They learn best by being prodded to push the envelope, and by being encouraged to undertake stimulating new challenges.

Mercury in Taurus When placed in the **fixed earth** sign of Taurus, the fleet-footed messenger moves at a slower pace. These are the patient deliberators of the zodiac who, despite the rhythms of those around them, work at their own pace and in their own way to process the information they are receiving. Though they can be maddeningly stubborn, fixating on a thought or an opinion, Mercury in Taurus is also dependable and stable—a grounded thinker unlikely to be swayed by the latest fad or fashion. Like the bull in the field contentedly grazing, however, Mercury Tauruses can suddenly charge—delivering their opinions with great power and force. Concrete and reliable, those with Mercury in Taurus may express Mercury's talent for trade and finance, excelling in economics and business. But Taurus is also a sensuous, artistic sign, with an inborn talent for art and music. Thus those born with Mercury in Taurus may flourish as singers, musicians, chefs, collectors of art, or connoisseurs of music, cuisine, and the finer things in life. Mercury in Taurus learns best through visual imagery, tactile, hands-on learning experiences, and by being left alone to learn in her own good time.

Mercury in Gemini Mercury in the *mutable air* sign of Gemini happily flexes its mental muscles. Here Mercury takes flight, soaring on wings of thought across vast tracts of knowledge, able to scale in a single leap intellectually complex concepts and notions. Ideas are food for the soul of Mercury in Gemini; his spirit is nourished on news, gossip, novels, magazines, lectures, and conferences. Mercury Geminis thrive in atmospheres bubbling and swirling with thought, talk, and speculation. Mercurys in this cognitive sign think purely for the sake of thinking, making them gifted writers, speakers, and facile communicators. Getting an idea across skillfully and clearly—any idea—is Mercury in Gemini's forte. Easily bored and distracted, however, Mercury Geminis must guard against skimming too lightly through life; their charming way with words can actually lead them away from the very truth they seek, while their ability to see both sides of everything can keep them suspended in intellectual limbo. Mercury in Gemini learns best by being kept constantly stimulated through an infusion of new, colorful, and intriguing ideas.

Mercury in Cancer In the *cardinal water* sign of the crab, the thinking processes of Mercury take on an intuitive, emotional tone. As it has often been said, Mercury Cancers think with their hearts, and learn through their feelings. These are the deeply private thinkers of the zodiac, those who hold their thoughts close to the chest until they feel they are in a secure-enough environment to express them. Sensitive, moody, and empathic, Mercury Cancers easily absorb the emotions subjectively coloring their surrounding environment that, in turn, affect their own positive or negative frame of mind. Gifted in the area of "emotional intelligence," Mercury Cancers read human nature the way others read books. Possessed of remarkable memories, they are experts at recollecting and re-creating the past. Still, they must guard against letting their thoughts drown in the watery world of emotion, and must strive to maintain a measure of objectivity in their communications. Cancer is a pioneering sign, and in their own introverted, crabwise way, those with Mercury in Cancer are busy blazing their own trails. Possessed of good instincts, and concerned with issues around family and survival, Mercury Cancers are clever money managers. Like Mercury Tauruses, food and security not only interest them, but may become outlets for careers as chefs or salespeople. Their talent for reading human nature enables them to excel as psychologists or novelists. Mercury Cancers learn best when they trust and like their teachers, and are provided with a secure and nurturing learning environment.

Mercury in Leo In the *fixed fire* sign of Leo, Mercury shines like a bright sun. Mercury Leos are the positive thinkers of the zodiac. Their thoughts are like warm flames on a cold winter day that bring cheer to the lonely and oppressed. Warmhearted and spirited, Mercury Leos can be dazzling speakers, or persuasive in any kind of public affairs. Their words inspire, give faith, and radiate courage. Mentally powerful, Mercury Leos' thoughts center on creativity and inspiration—the "gold" of imagination. But Mercury Leos like real gold, as well, and may have the Midas touch when it comes to business and commerce. Like the lion or the royal king or queen, two symbols attached to the sign of Leo, Mercurys in this sign will communicate with dramatic flair, pride, and self-confidence, and their manner will be innately regal. Hardly the shrinking violets of the zodiac, they do well in the center of the spotlight, making excellent actors, artists, and speakers. Yet Mercury Leos must watch for excess pride and the dangers of grandiosity and inflation. Though they must seek the honor their bright minds deserve, they need to guard against self-aggrandizement or boastfulness. Mercury in Leo learns best when given a starring role, or a position of authority where her mental powers are allowed to beam brightly.

Mercury in Virgo Mercury in the **mutable earth** sign of Virgo is the position of the analysts and practical thinkers of the zodiac who take pleasure in life's little details. Those born with Mercury in the sign of Virgo possess sharp powers of mental discrimination and are skillful in the art of logic. Gifted with the ability to find the pattern in the pieces, Mercury Virgos make excellent artists, writers, nutritionists, healers, accountants, or computer programmers. Health and the improvement of the function of the body are also matters of great interest for those with Mercury in Virgo. Because they are duty and service oriented, their thoughts may revolve around work and responsibility. Most important, Mercury Virgos give much thought to *understanding*—they simply want to know how and why things are the way they are. Indeed, unless focused on a project that employs his prodigious faculties of discrimination, Mercury Virgo's mind can mentally spin out in overattention to useless and meaningless details. At his best, Mercury Virgo is able to break things down into bite-size pieces, carefully examining every facet and angle of a subject until arriving at a well-thought-out conclusion based on facts and evidence. Those with Mercury in Virgo must guard against being overly critical and intellectual, or too coldly perfectionistic. Obsessed with examining every little blade of grass or tree branch, they all too often miss the "big picture," or the forest for the trees. Mercury Virgos learn best in an environment that is orderly and well structured, and where all the facts are presented to them in a down-to-earth and logical manner.

Mercury in Libra In the **cardinal air** sign of Libra, Mercury's mental outlook is one of balance, fairness, and repose. Placed in the sign that rules relationships, Mercury Libras are adept at dialogue and conversation. Innately gifted with the ability to see the other person's point of view, those with this placement tend to make excellent diplomats, teachers, counselors, or marriage therapists. Rarely does their thinking process occur in a vacuum, but rather it occurs in a context of exchange and a flow of information. Their thoughts are formed in the vessel of relationship. The symbol of Libra is that of the scales, and Mercury Libras give great thought to issues of equality and rights. Mental disturbance results when life seems out of balance, or when a situation is being handled unfairly. Possessed of rational intellects, these are the "legal eagles" who bring strict neutrality to their social dealings. The higher octave of Mercury in Libra is that of bringing beauty and harmony—finding and saying just the right words—where communication has turned sharp or unpleasant. These are the travelers of the "middle road." On the other hand, Mercury Libras must watch for the tendency to vacillate between opposing points of view, losing their own standpoint in the process. Indecision, or hanging in the balance, as well as an avoidance of emotional confrontation, are other dangers. Mercury in Libra learns best in an atmosphere that is serene and undisturbed; and where knowledge is presented in a clear, unbiased, and balanced fashion.

Mercury in Scorpio Those with Mercury placed in the *fixed water* sign of Scorpio possess a frame of mind that is passionate, private, and intensely interested in that which is hidden. Drawn down to the depths, Mercury Scorpios are obsessed with thoughts of solving the riddles of life. They are the detectives, the mystery writers, the occultists, or the muckraking journalists whose task it is to probe beneath the surface. With little interest in subjects of a shallow or fleeting nature, Mercury Scorpios are mental deep-sea divers. Diving deep and surfacing, they return with a treasure others may have overlooked. Innately psychologically oriented, Mercury Scorpios are fascinated by the unconscious and the unseen. Though she may not be a psychologist by profession, you can be certain that a person with

Mercury in Scorpio is furiously computing the hidden motives and drives of every person with whom she comes in contact. Few secrets remain hidden for long from the prying thoughts of Mercury Scorpios: they are relentless in their search for the key to just about everything that exists. The highest octave of Mercury Scorpio is Hermes in his role as psychopomp, or guide of souls between life and death. While they may probe the mysteries of life, however, Mercury Scorpios can be cautious and withholding, possessively keeping their thoughts to themselves. Those with this placement also may find themselves struggling with negative thoughts of mistrust, resentment, and paranoia. Mercury in Scorpio learns best in an environment where she is allowed to enter deeply into a subject, whether it be history, forensics, or psychology.

Mercury in Sagittarius In the *mutable fire* sign of Sagittarius, mental Mercury roams far and wide throughout the world of knowledge, a seeker in search of truth and wisdom. Those with this placement have their minds set on the distant horizon, where high ideals and the future beckon them ever onward. Possessed of restless, questing minds, those with Mercury in Sagittarius are visionaries who can inflame, inspire, and excite others through their contagious ideas. These are the preachers, the teachers, the publishers, the editors, the New Age counselors, or the television talking heads. Indeed, Mercury Sagittariuses are at home in their heads, curled up in a mental milieu of ideas, the more expansive and far-reaching the better. Unconcerned with detail, these are the "big picture" thinkers, those who are gifted at connecting the dots. Holistic, right brained, synthetic, process oriented, Mercury Sagittarius intuitively grasps the grand design, the pattern in the stars. The fire of their minds is fed by the fuel of intellectual exchange; lively conversationalists, broad-minded Mercury Sagittariuses never met an idea they didn't like or a concept they couldn't explore more fully. Those with this placement must guard against a tendency toward "high-mindedness," or allowing their mental preoccupations to lift them too far off the ground of everyday life. Sagittarius Mercury learns best in an environment that stimulates his intellect with a broad and fascinating array of topics, and where he is never allowed to be bored.

Mercury in Capricorn Those who have their Mercury in the *cardinal earth* sign of Capricorn are possessed of a penetrating, steely intellect that brooks no fools. Capable of being both pragmatic and deeply thoughtful, these are the serious thinkers of the zodiac whose minds are occupied with weighty matters of survival and success. With their intuitive grasp of business and financial strategy, Mercury Capricorns make excellent CEOs, economists, business owners, or captains of industry. The symbol of Capricorn is the mountain goat; thus those with this placement are possessed of a frame of mind that is patient, persistent, and doggedly determined to get to the top of the mountain. A Mercury Capricorn does not bear mental confusion well but will organize her thoughts methodically, in order of priority. Naturally politic with a talent for politics, Mercury in Capricorn has an innate grasp of how the world works—as well as how to bend it to her own will. But Mercury Capricorns can be scholarly and academic minded, as well. Lone thinkers with a tendency toward reclusiveness and solitude, they might shut themselves away in a library, only to emerge years later with a tome on philosophy or a history of economics. Cautious, reserved communicators, those with Mercury in Capricorn tend toward melancholy and depression, and must guard against excessive worrying, cynicism, and skepticism. Yet on the other hand, Mercury Capricorns are the ironists of the zodiac, with a sharply honed dry wit. Mercury Capricorns learn best in an environment that is well-organized, goal oriented, and neatly structured, and where even the most abstract ideas are grounded in everyday reality.

Mercury in Aquarius In the *fixed air* sign of Aquarius, Mercury breathes free, soaring high above the earth on currents of abstract thought. These are the Utopian dreamers of the zodiac—humanitarians whose thoughts center on improving the human condition. The stable nature of this air sign anchors their ideas and ideals in reality, however, allowing the possibilities they dream of to take shape in the real world. This makes Mercury Aquarians excellent inventors, scientists, theoreticians, or political revolutionaries. Though eccentric individualists, the thoughts of those with Mercury in Aquarius rarely center on themselves alone, but on humanity as a whole. Unless continuously nourished by contact with a variety of people, foreign cultures, ideas, and systems of thought, the mind of a Mercury Aquarius starts to suffocate. Rarely do those with Mercury in Aquarius find themselves fixating on the past; rather, their frame of mind is future oriented, constantly evolving toward new horizons. Detached and objective, their communications style is cool, clear, and constructive. Mercury Aquarians tumble from their ivory towers when they encounter the irrational realm of the emotions, however, where their rational ways of thinking do not serve them very well. Mercury Aquarians learn best in an environment that engages their mental powers in an evolutionary and inventive manner, where they can begin to creatively apply their brilliant solutions to the human condition.

Mercury in Pisces In the *mutable water* sign of Pisces we meet the magic thinkers of the zodiac. At home in the deep waters of the imagination, those with Pisces in Mercury are poets and mystics, artists and psychics, filmmakers and photographers. Intensely emotional and subjective, Mercury in Pisces' thoughts are colored by the mental static that drifts in from all realms and dimensions, making it difficult to distinguish which thoughts are their own—and which belong to those around them. Impressionable as the still glass of the surface of a lake, Mercury Pisces are empaths, sensitive to the emotional atmosphere within which their minds are constantly swimming. The mind of a person with Mercury in Pisces works, not in a logical or rational way, but through intuition and osmosis. They think in images, myths, and metaphor, and may communicate their thoughts in a nebulous and stream-of-consciousness manner. Devoted to the

5.

THE ASTRONOMY
AND ASTROLOGY OF
MERCURY RETROGRADE

> I think that any retrograde planet tends to interiorise
> the expression of the planet. It operates on a more
> subjective and covert level. The meaning of a planet
> doesn't change, but the capacity for extroverted
> expression is altered.
>
> —Liz Greene, *The Outer Planets and Their Cycles*

What happens when Mercury, the planet of communications, turns retrograde? How will its change in motion affect you? What does it mean if you were born during Mercury retrograde, or if your calendar birthday falls during this time period?

If you have ever experienced the disruptions caused by a blizzard or a severe thunderstorm, you will have some idea of the haphazard effects that can be stirred up when Mercury enters its retrograde phase. When a major storm hits, for example, your electricity may go out. Suddenly, you can't work at your computer, make cordless phone calls, microwave popcorn, or watch television. Drifts of snow, fallen trees, or downed electrical wires make an ordinary trip to the grocery store a major trek. Whatever projects you were working on have to be put aside as you find yourself preoccupied with unexpected challenges: digging out your car, draining a flooded basement, or taking care of bored children home from school.

If the storm's effects last over several days, or, in some cases, even weeks, you begin to find yourself adapting to your changed routine. Instead of impatiently railing against the lack of modern-day conveniences, you enjoy such newfound pleasures as reading a book by candlelight, or sharing stories with your children in the dark. Even the additional physical exertion expended in clearing snow or debris or making repairs may begin to feel rejuvenating. Then, one day, electricity is restored. The snow melts, the streets are cleared, and life returns to its normal busy pace. Yet somehow, your encounter with nature's vicissitudes has changed you. Perhaps you feel restored by a sense of inner calm, a renewed connection to nature, or deepened by your intimate connection with loved ones. As a result, your perspective on life has shifted, and what once seemed so important has been replaced by a new set of possibilities.

In other words, as the metaphor of a storm shows, Mercury retrograde marks a significant period of transition when life-as-we-know-it may feel random, fickle, and out of sync. During this three-week interval, events may occur that will shift the focus of our attention in a different direction. Just as we have no choice when a storm hits but to drop what we've been doing and pick up a shovel or eat cold food from a can, so, too, during Mercury retrograde must we adapt to a change in rhythm. Far from being "bad," however, this transitional shift may actually serve to correct our life course, putting us back on a track that is more aligned with our deeper selves.

Like a piece of music with variations of rhythm, even this period is characterized by changes in tempo that range from upheaval and confusion to solitude and mental clarity. As astrologers know, each twenty-one-day cycle is marked by several important phases based on how quickly or slowly the planet appears to be moving in the sky and its position relative to the Earth and Sun. Indeed, the first step in understanding the dynamics of how Mercury retrograde works begins with the Sun. "The retrogradation of a planet," says scholar and Hellenistic astrologer Robert Schmidt, "is only one moment in a much larger cycle of that planet relative to the Sun." The center of the solar system around which all the planets whirl, the Sun's annual cycle through the sky marks our seasons. Psychologically, the solar principle represents the ego, or one's empowerment to act as an individual, as well as the masculine, positive force of life. Like a planetary diplomat, Mercury negotiates the boundaries between the solar self and the world, translating raw will into intelligible action. Aided by Mercury's communications' skills, the Sun carries out its ambitious plans, expanding, building, and accomplishing its heroic aims.

When Mercury is direct, the solar force of life streams outward in an unimpeded and forthright fashion. During this time, writes Erin Sullivan in *Retrograde Planets*, "there is little time for retrospection, and energy is expended in productive action. Our contemplative side is virtually on hold." When the planet Mercury slows to a stop and begins to retrograde, however, the usual forward-moving drive to get things done and make things happen is interrupted. The period of time that Mercury is retrograde, writes Sullivan, can be described as a mental "downtime." This period presents us with the golden opportunity to look back over the experiences of the previous three months, and to process and digest our feelings. It allows us to review potential options for the future, giving us the time we need to realign with our true self, and return once again to the path of action that more accurately reflects who we really are.

These two alternating rhythms, explains Sullivan—Mercury direct and Mercury retrograde—are as natural as waking and sleeping. Maintaining a balance between repose and activity, or between being and doing, is as necessary to our physical health and mental well-being as eating well and exercising . In *A New Look at Mercury Retrograde*, astrologer Robert Wilkinson likens the Mercury retrograde-direct process to a gradual "in-breathing and out-breathing. The motion (of Mercury) gradually decreases in speed to a point of maximum slowness, then gradually increases in speed again over time." To gain the maximum effect of Mercury retrograde, however, it helps to understand the unique rhythm and meaning of each of its phases.

THE THREE PHASES OF
MERCURY RETROGRADE

The first phase of Mercury retrograde, known as the **stationary retrograde** period, occurs approximately four or five days before, and several days after, the exact hour when Mercury appears to stop and move backward against the sky. From an astronomical perspective, Mercury is farthest from the Sun in longitude and is about to swing back toward the Earth. This means it has just reached its greatest "Eastern Elongation" and has appeared on the horizon at sunset as the evening star. As Mercury stations to a halt, abruptly impeding the usual forward motion of life, there is a dramatic and disruptive clash of rhythms and emotions can run high. During the heightened interval that occurs at this initial stage, astrologers advise greater caution and awareness, as confusion and the potential for mistakes are even more pronounced. Blocked by circumstances or taken aback by unforeseen developments, one cannot move forward, but only backward or inward. "The degree to which we force results," writes Sullivan, "is the degree to which the initial phase of the retrograde cycle is frustrating."

Approximately ten days after the onset of the stationary retrograde phase, Mercury enters the second phase of its retrograde cycle, or the **inferior conjunction.** At this critical juncture, the planet will reach its closest point to the Earth, lining up between the Sun and the Earth. From the terrestrial standpoint, a "conjunction"—or coming together in the same degree of an astrological sign—of the Sun and Mercury appears to take place in the sky. The day this occurs is also known as the "New Mercury," and marks the **midpoint** of the Mercury retrograde cycle. The week following this conjunction often signifies a turning point, as the clouds of confusion and disarray disperse and the pace of life may begin to quicken. At this stage, one gets a "vague sense of what is to come," writes Wilkinson, "but no clear picture as yet." Thus while it may be a good time to make plans and plant seeds for the future, astrologers continue to warn against implementing any final course of action until Mercury is finally direct. By resisting the impulse to act or make major decisions, we can use the time to further consolidate the inner work that has been taking place below ground.

After the inferior conjunction, Mercury begins traveling away from the Earth, reaching its farthest point from the Sun, or "Western Elongation," rising on the horizon as the morning star. On the cusp of moving fully direct again, Mercury now enters its third phase, the **stationary direct** period, where it once again hovers for four or five days before, and several days after it has gone direct in the specific degree of a sign. During the charged, powerful days of this threshold station, a feeling of release and relief pervades the atmosphere. Many things can begin to happen at once, and the atmosphere may feel electric with mental anticipation and renewed activity. What has been scattered falls into place, decisions are reached, clarity of mind dawns as what has been hidden emerges into plain view, and projects receive the green light. As Mercury moves forward so, too, can we go forward with our plans, unhampered by introspection.

When Mercury may officially go "direct" and move forward in the skies, many astrologers still advise their clients to make as slow and conscious a reentry as possible. For one thing, they say, it will take several more weeks before Mercury, even though it is moving quickly in the skies, will reach that degree in the zodiac where it first stationed retrograde—this is called the **shadow period.**

The **shadow period** can be explained this way. When Mercury turned retrograde on December 17, 2003, for example, it had reached 12 degrees of Capricorn along the zodiac wheel. When it stationed direct three weeks later on January 6, 2004, it had moved backward in the zodiac until it had reached 26 degrees of Sagittarius. Over the next three weeks, Mercury moved forward, once again covering the same degrees along the zodiac wheel, until reaching 12 degrees of Capricorn on January 25—the same degree at which it initially stationed retrograde. Thus the three-week period that it takes Mercury to catch up to its original point in the zodiac before it turns retrograde is known as the *shadow period*. At the conclusion of the shadow period, Mercury has finally completed the territory it has covered—and re-covered—during the retrograde cycle. Astrologically, the significance of the shadow period indicates that although Mercury may be "officially" direct, it may take another three weeks until those affairs that were disrupted under the influence of Mercury retrograde fully return to normal or reach clarity. Likewise, a similar *shadow period* is in effect approximately two weeks before the day Mercury goes retrograde. On November 30, 2003, for instance, Mercury reached 26 degrees of Sagittarius—the exact degree at which it would station retrograde nearly five weeks later on January 6, 2004. Thus the *shadow periods* preceding and following the twenty-one days of Mercury retrograde could be said to be like a wave that gradually rises, crests, then falls, melting back into the ocean.

Technically, the Mercury cycle is said to begin with the *inferior conjunction*—when the Sun and Mercury conjoin during the midpoint of the retrograde cycle—and peak at the *superior conjunction,* or the joining of the Sun and Mercury that occurs about two months after Mercury has gone direct. The *superior conjunction* is like the "Full Mercury," when the seeds that were planted at the New Mercury, or *inferior conjunction,* reach fruition. The forty-seven days between the Full Mercury and the next stationary-retrograde, explains Sullivan, are when action is completed and rewards are reaped—then the cycle repeats itself again. For those who are sky watchers, it is interesting to note that Mercury is visible only during its greatest elongation points, when it is farthest from the Sun and appearing either as the evening or morning star.

Combining the astrological and astronomical symbolism of Mercury retrograde in a way that makes sense may seem a bit confusing at first. But scholar Robert Schmidt, who has translated centuries-old Greek manuscripts on astrology, offers a useful metaphor for understanding the everyday, practical effects of Mercury retrograde. The

Greek word for retrograde, he says, literally means "to walk backward." Metaphorically, he says, it also means to recall something back, like when an automaker puts out a defective vehicle and the manufacturer calls it back. Thus when Mercury is retrograde, continues Schmidt, it operates as a "planet of contest"—just as in a lawsuit, when cases are contested and then overturned. Mundane affairs undergo a period of suspension and uncertainty; projects are left dangling or unfinished. Fortunately, this state of affairs is relatively brief, lasting for approximately three weeks when Mercury emerges from retreat and resumes normal functions—and a verdict, or an outcome, is eventually reached.

MERCURY AS COSMIC INSTRUCTOR: LEARNING FROM THE PATTERNS OF MERCURY RETROGRADE

From an astrological perspective, Mercury retrograde is a collective phenomenon that impacts—or disrupts!—the whole Earth. Thus the sign Mercury is in at the time of its retrograde phase will shape the cycle in a particularly expressive way. Further, the sign that Mercury is in will impact each person's chart in a slightly different manner. The natal chart of an individual, for instance, contains a unique arrangement of the zodiacal signs in the pie-shaped twelve houses: one person may have an Aries first house, for instance, while another may have an Aries third house. Likewise, one person may have a Gemini tenth house, while another may have a Gemini twelfth house. Thus when Mercury retrogrades in the sign of Aries, everyone will experience Mercury through the lens of that sign—yet for the individual it will play out in a specific house, or area of life (a glossary of astrological terms and symbols appears at the back of the book and includes illustrations that explains this phenomenon more clearly).

Even more interesting, however, the retrograde phases of Mercury actually trace a *pattern*. Over the course of approximately thirteen months, Mercury retrograde traverses the three signs of each of the four elements tracing a trigon, or grand triangle, in the process. As Erin Sullivan explains in *Retrograde Planets,* there is a "gradual precession of Mercury backwards through the elements" moving from Water, to Air, to Earth, to Fire. In contemplating this unusual pattern, Sullivan recalls Hermes' cattle theft as a precocious child, when he reversed their hooves in order to escape Apollo's detection. Yet just as Hermes' theft led to Apollo's gifts of the lyre in exchange for prophecy, so, too, do Mercury's tracks lead in the end to the treasure of insight and knowledge.

Following Mercury's backward trail through the twelve signs, we follow the soul guide as he inscribes a message in the sky. In these cycles of Mercury retrograde through the signs and elements, we witness the planet as a kind of celestial instructor delivering his "life lessons" from the cosmic book on life. At the time of Mercury's stationary retrograde, as just mentioned, the planet has just traveled from around the Sun, and is approaching its nearest point to Earth. Thus as many astrologers have pointed out, Mercury in its retrograde phase could be said to be delivering a message directly from the Sun, the center of our solar system, to those of us on Earth.

All the individual interpretations for Mercury through the signs in the previous chapter can be used to further illuminate the influence of Mercury retrograde. Each of the four elements, however, also offers us a set of guidelines for what to expect. Mercury retrograde in the air signs of **Gemini, Libra,** and **Aquarius,** for instance, may be imparting wisdom in the field of communications, relationships, and politics. During this time we may need to cultivate mindfulness and consciousness when communicating our thoughts to co-workers or loved ones. Or, we may use this period to reexamine some of our long-held assumptions and beliefs, discovering that our usual frame of mind has become outworn or is counterproductive. It may also be a time to deepen and intensify our mental concentration. In the earth signs of **Taurus, Virgo,** and **Capricorn,** Mercury may be asking us to review how we handle our resources. The practical details of life—

money, possessions, food, and housing—may demand more of our attention. In the process of attending to these everyday matters, we may discover the need for a new set of values more in harmony with our ideals, reconstructing the foundation upon which our lives have been built.

In the fire signs of **Aries, Leo,** and **Sagittarius,** Mercury retrograde fans the flames of our creativity and childlike spontaneity. Here we are being asked to reexamine where we have become *too* conventional and practical minded. Where, asks Mercury, have we snuffed out the fires of our passion, ignored the intuitive impulse to forge new ground and travel toward new horizons, or dimmed the playful side of our natures? Finally, when Mercury is retrograde in the water signs of **Scorpio, Cancer,** and **Pisces,** we are called to plumb the depths of our emotions, examining psychological patterns and becoming resensitized to the hidden, watery world of our feelings. Matters of the heart, and the areas of love and relationship may draw our attention. As the water signs are reflective and empathic, Mercury retrograde through these signs may also be a time to study the psychic arts; dream work, astrology, tarot, and increasing telepathic sensitivity are favored during these times.

BORN UNDER MERCURY RETROGRADE

The fact that Mercury retrograde occurs three times each year means, of course, that a significant number people are born with Mercury retrograde in their astrological birth chart. If you have a copy of your chart, this is indicated by the letter *R* placed in type beside the symbol of the planet Mercury: ☿ ℞ (A timetable of periods that lists the year and month when Mercury is retrograde and direct, as well as the sign it is in, is included in appendix C).

Many of the archetypes and attributes that have been used to describe Mercury retrograde offer insight into how this placement functions in an individual's chart. Most significantly, just as the three-week cycle of Mercury retrograde signals a period of introspection, a person born with Mercury retrograde will tend to be more deliberately thoughtful and introverted by nature. A retrograde Mercury, writes Howard Sasportas in *The Inner Planets,* "inclines you to look inside yourself, perhaps to check out what you're going to say before you speak, and it also inclines you to chew over any thought or idea you have . . ."

Lynn Koiner, who has practiced and taught astrology for over thirty years, says that those individuals born during Mercury retrograde are rarely superficial. Rather, she says, a natal Mercury retrograde often indicates a deep and profound thinker who repeatedly analyzes situations in his attempt to figure things out. Like mythic Mercury's role as mediator between life and death, these individuals feel compelled to dive to the bottom of things in search of the pearl of knowledge and insight. Far from weakening the mind, writes Erin Sullivan in *Retrograde Planets,* Mercury retrograde powerfully stimulates a person's reasoning abilities. Driven to understand as much as she can, a person born with Mercury retrograde may have an inborn—even insatiable—talent for research or any kind of detective work. Indeed, writes Sullivan, most born with Mercury retrograde have "great mental endurance and the stamina for long and arduous projects involving research and inquiry." Koiner notes that Mercury retrograde can be the signature of the scientific thinker; many astronauts born during the late sixties and early seventies, she says, had Mercury retrograde. Astrologer Laurence Hillman adds that, in his experience, a person born with Mercury retrograde possesses a mind that "will develop counter to whatever the going norm is. In addition, they may unfold later in life."

There can, however, be a downside to being born under Mercury retrograde. For the most part, the negative side of this aspect stems from an overexaggeration of thinking and reasoning. The tendency to overanalyze and logically sort everything out can lead to mental exhaustion, says Sullivan. Likewise, the Mercury retrograde individual's habit of gathering as many facts and as much information as possible before making up his mind can lead to anxiety and chronic indecision. Perhaps because of their natural interest in the deeper side of life, many who have come to me for consultation have Mercury retrograde in their natal charts. Among other characteristics, I have noticed that those with Mercury retrograde have a pronounced tendency to think things through almost to an extreme, taking in as much information as they can before reaching a conclusion. Although this means they have patience and thoughtfulness of character, it also means that they must prod themselves to take the risks that life sometimes requires. In their quest to have all the facts, points out Robert Wilkinson in *A New Look at Mercury Retrograde,* people with Mercury retrograde often attempt to "take too much into consideration"—a tendency that may lead them down sidetracks or detours in the interest of exploring every available option before reaching a final decision. Indeed, Mercury retrograde individuals may sometimes feel as if they are always traveling toward a destination that continually recedes into the distance. Like Odysseus, they may think they are headed in a specific direction, then abruptly find themselves in a strange and unexpected place. Those who are under the influence of Mercury retrograde may feel like wanderers, searching for an elusive homeland.

Those with Mercury retrograde, in fact, can often seem to others as if they are always going off on tangents, developing obsessions with hobbies or pursuits. One woman I know with a retrograde Mercury, for instance, dropped a promising career in the art world and became a massage therapist—while honing her talent in needlework and embroidery. Just as Mercury embodies the twinlike qualities of duplicity and duality, so Mercury retrograde people will often have more than one interest. In addition, those born with Mercury retrograde can sometimes seem dreamy and preoccupied. Because their thinking function runs counter to the rest of the world, they are busy musing and reflecting, and their attention may often seem far removed from the present moment. Their thoughts, says Lynn Koiner, can become highly subjective and "some may become so absorbed in their own internalized process that they seem absentminded and unaware of external activities."

Indeed, though Mercury retrograde does not automatically mean difficulties in learning or communications, it does indicate someone whose communication style and learning process may differ from those in the mainstream. Children born with Mercury retrograde, for example, may suffer in school—not because they are less intelligent, but because they move at a slower pace than their peers. Astrologer Ray Grasse says that the tendency of children with Mercury retrograde in their charts to hesitate or be reticent can cause them to feel stupid—which they aren't. "It's the proverbial kid in the classroom," he says, "who knows the answer before anyone else, yet raises their hand last; eventually, they may go on to become the head of the class."

In a culture that equates fast with smart, however, the slow-thinking child can sometimes be diagnosed with ADD or other learning disabilities. This was the case with one of my Mercury retrograde sons who, although placed in a program for gifted and talented students, felt pressured by the demands of the academically accelerated classes he was placed in. Accommodating his needs, his father and I placed him in a regular class where he was able to learn in a less stressed and hurried manner. As astrologer Robert Wilkinson writes, Mercury retrograde children learn in a "roundabout" process and must be encouraged to study "in their own way, at their own pace, within reason." Mercury retrograde minds, he says, do better when they are allowed to explore many possibilities at their own rhythm, as they have a talent for unusual points of view and for finding value in what others have overlooked. Because society tends to place more value on regimentation than originality, however, points out Wilkinson, people born with Mercury retrograde can seem out of step with the culture at large.

When it comes to expressing their innermost emotional feelings, Mercury retrograde individuals will be more likely to take the indirect, nonconfrontational approach. Often, says Lynn Koiner, this is because as children they may have felt that no one listened to them, or that they had no one to share their thoughts with. One man I know, who was born with Mercury retrograde in the conversational sign of Gemini, had no problem waxing eloquent on topics such as politics, movies, or books. But when it came to emotional intimacy, he told his partner that she needed to "set the table," so to speak, before engaging in personal issues—in other words, he needed advance warning in order to prepare himself before he could be forthcoming with his feelings.

Koiner says wisely that this is because the person born with Mercury retrograde needs time to "brood and ponder." Mercury retrograde people need time to digest and process information, says Koiner, as they are often "slow to defend themselves, and slow to respond to queries of a personal nature unless they are prepared ahead of time." Thus if you want to discuss a personal matter with a Mercury retrograde friend or lover, she advises, "you must tell them what you want to discuss and set up an appointment to discuss it. If you want immediate feedback, you will not get an honest or well-thought-out answer," she says, as they don't immediately know how they feel about something. In addition, they appreciate a slow, measured response in others.

Although the Mercury retrograde person may be slow to express her feelings, or to respond to someone else's emotions, she is highly perceptive and intuitive. Beneath her surface of apparent calm, she is busy reading the nonverbal cues of those around her. Possessed of sensitive psychic antennae, she assesses the body language and facial expressions of people she is interacting with, interpreting the invisible emotional currents that are being communicated without words. For this person, communication occurs as much in the silence between words, as in the words themselves.

In a culture that runs on ever faster and faster rhythms, the unique challenge posed to the person born with Mercury retrograde is to allow himself the time and space to discern what it is that he has been called to "figure out." What is the issue, problem, or project that is so strongly compelling his attention elsewhere? Whether he is grappling with solving a scientific puzzle, an unconscious dilemma, or an overwhelming interest in a particular field of study, the task of the Mercury retrograde thinker is to give himself permission to follow his muse—in whatever direction it takes him, and at as slow and steady a pace as he wishes. For at some point, whatever it is he has been processing in the hidden recesses of his deepest self will someday come to light. In the words of the famous poet T. S. Eliot, after all their adventures they will arrive at the place from which they first began, and "know the place for the first time."

In fact, the most important thing for those born under a Mercury retrograde to remember is that while this astrological aspect indelibly imprints a person's thinking and communications style, at some point they will undergo a significant shift. Just as it does during its thrice-yearly cycles, Mercury will turn direct in an individual's chart through a process astrologers call "progression." This means that by the age of twenty-one or before, the Mercury retrograde person will experience a release of mental energy that has been held back since birth. Like water released from a dam, his thoughts will become more free-flowing and expressive. As Howard Sasportas writes of this time, "They . . . come out of themselves more, opening their minds to new things and generally becoming more extroverted."

For example, says Ray Grasse, take the brilliant New Age thinker Ken Wilber, who was born with Mercury retrograde in Aquarius. A shut-in for many years who holed up with his books and writing, Wilber, says Grasse, "has come out of his shell in recent years in a way that has surprised many of his colleagues." In other words, explains Grasse, "things can turn around with this pattern, and almost go to the other extreme."

At the same time, an individual who was not born with Mercury retrograde may experience this phenomenon later in life. If Mercury is retrograde on the day of your calendar birthday, for instance, you will experience its effects for the following year; it will begin to wane several months before your next birthday. In addition, just as Mercury may turn direct by progression, as described above, so it can also turn retrograde during adulthood. This is a period that lasts for twenty-one years and marks a significant life period of inward self-reflection, study, and deep research during which one's perspective on life may undergo a radical change in orientation. Indeed, a person who undergoes Mercury retrograde by progression this way will be called upon to develop his own voice and to become an original and independent thinker. (For those who are interested, a trained astrologer can tell you whether or not you will experience a "progressed Mercury retrograde" during your lifetime.)

In order to understand the way Mercury retrograde influences your life more specifically, it helps to observe its effects in the particular house—or sphere of activities—and astrological sign that it falls in. The following is a brief outline of Mercury retrograde through each of the signs and twelve houses of the zodiac. Here is how this tool of interpretation works: Each of the twelve houses in the circular chart is "ruled" by one of the signs. These signs are the key to understanding the meaning of each house. Aries rules the first house, Taurus the second house, and so on. Thus if you know that Mercury is currently moving retrograde in the sign of Cancer during one of its three annual cycles, you would read the section marked "Mercury retrograde in Cancer and the fourth house" for what to expect.

If you have a copy of your birth chart, you will have the opportunity for an even more detailed interpretation of how Mercury retrograde is impacting you individually. (If you do not already have a birth chart, you can find several fine Web sites in the Resources section where you can order or download one.) Though it may sound confusing at first, each birth chart is unique because according to the time you were born, the twelve signs fall in different houses (though the houses still derive their meaning through their original sign rulerships). For example, if the sign of Cancer is in the fifth house in your birth chart, then you would read the interpretation for *both* Mercury in the fifth house *and* Mercury in the fourth house in Cancer, then combine their meanings. The same set of guidelines applies if Mercury retrograde falls on your birthday; this time, however, the influence lasts until about three months before your next birthday. This technique is also true for the progressed chart, only the influence lasts for a much longer period of approximately twenty-one years.

MERCURY RETROGRADE THROUGH THE HOUSES AND SIGNS

Mercury retrograde in Aries and the first house of selfhood and initiative. Mercury retrograde in Aries and the first house stirs issues around self-assertion, identity, how you appear, and how you interact with others. Rather than the usual Aries-style "take-the-bull-by-the-horns" approach, you may feel more like sitting back and thinking before you act. Thus Mercury retrograde in this position may mean that you delay any kind of action that requires your personal initiative and impetus. Rather than leap into a new enterprise or instigate any rash confrontations, it is a time to hold back and adopt a more circumspect approach. On a deeper level, this is because you are preoccupied with that age-old question "Who am I?" and reevaluating whether your personality style is in alignment with your authentic core self. You may even experiment with changes in your style of dress and and the way you wear your hair. When Mercury begins to move forward, you will feel a renewed sense of selfhood, as well as more confidence and courage.

Mercury retrograde in Taurus and the second house of finances, resources, and values. When Mercury is retrograde in Taurus and the second house, it is a time to be especially cautious in financial matters. Double-check bank statements and credit-card charges. Hold off on any major changes in your investment portfolio; analyze different options, but don't sign any papers or move any money until

Mercury is direct. On a deeper level, Mercury retrograde in this house and sign may signal a period of introspection around the values you hold; you may reexamine what really gives you security, as well as the moral foundation upon which you have built your life. While some may discover that they have neglected the practical side of life in favor of spiritual or creative pursuits, others may realize that they have focused on the material things in life to the exclusion of other kinds of meaningful activities. Mercury retrograde in Taurus and the second house can also signal a time to slow down and taste the more physical and sensual sides of life: leisurely dinners with wine and good friends; gardening, going to concerts and listening to beautiful music; visiting a health spa; or taking walks in the woods.

Mercury retrograde in Gemini and the third house of communications, writing, and speaking. Mercury retrograde in Gemini and the third house signals a time to be especially mindful around all forms of communication—from personal conversations to letters and e-mails. According to Lynn Koiner, Mercury retrograde in this placement favors study and research in a quiet atmosphere where thoughts can be clearly formulated. Because the third house rules short trips and neighborhoods, the tempo of daily life may slow. You may find that you are drawn into more exchanges with neighbors or local merchants. It is also a favorable time to engage in conversations of a deeper, more thoughtful nature. Silence, as well, can be golden, as you take the time to communicate with your inner self. This may not be the best time to deliver lectures or public presentations that rely on statistics or facts; on the other hand, any kind of public speaking or teaching will benefit from Mercury retrograde's deeper perspectives or its talent for looking at life from an unexpected angle. As this is the house that rules siblings and neighbors, Mercury retrograde here may bring a visit or encounter with a relative. Journaling, catching up on back issues of newspapers and magazines, and other forms of creative writing are also favored at this time.

Mercury retrograde in Cancer and the fourth house of the home and family. Mercury retrograde in Cancer and the fourth house will turn your attention to the house you live in, your family life, and other domestic issues around safety and security. You may feel an increased need for a home environment that offers more seclusion and solitude. Or, you may become discontented with how your house is structured, and begin plans for renovation. Any rebuilding already in progress, however, may suffer delays and setbacks. Your parents may come to visit, or you may simply feel like staying at home more, recharging your batteries for when Mercury goes direct. Your notion of family itself may undergo a transformation, as you rethink the dynamics at play in your marriage or how you parent your children. You may delve into your family's past through genealogical research. On a deeper level, Mercury retrograde in this placement is symbolic of the roots of your personal unconscious. Thus it is a good time for any kind of psychological inner work, exploring your psyche and any unaddressed issues from childhood, whether in therapy or through journaling.

Mercury retrograde in Leo and the fifth house of creativity, pleasure, and children. As Robert Wilkinson writes in *A New Look at Mercury Retrograde,* "This is the time for a return of old playfulness . . . Here the inward-turned mind explores unusual symbolic art forms, or receives insight into games." Indeed, Mercury retrograde in the fifth house and Leo can be a magical time, if you let it be. The right-brained, intuitive way of thinking that is enhanced during Mercury retrograde finds a fertile field to play in in the fifth house. Thus it is a good time to paint, garden, sculpt, or experiment with a new recipe. If you are a teacher or a parent, it is especially favorable for reveling in the wonder of a child's world. As this is the house and sign placement of romance and love affairs, an old flame may return; or, you may find insight into your current relationship patterns by rereading old love letters or reminiscing about your romantic past.

Mercury retrograde in Virgo and the sixth house of everyday work routines, service, and health. Mercury retrograde in Virgo and the sixth house may bring issues in your working environment to the foreground. There may be disagreements and miscommunications among co-workers, or you may find that your normally punctual routine is overturned by unexpected disturbances. People may be more driven to perfectionism in their work, and, as a result, be forced to contend with issues around frustration and unrealistically high expectations. Health concerns may surface, or you may feel the need to completely reexamine your diet and exercise routines. Any decisions made at this time, however, might be subject to revision when Mercury goes direct. On a deeper level, Mercury retrograde in this segment of the zodiac signals a time to reflect upon your ingrained habits and how they impact your health and well-being—housecleaning away those old habits that are negative and counterproductive, and buffing up lifestyle routines that are strengthening and that cultivate efficiency. As Virgo and the sixth house rule the apprentice-teacher relationship, as well as mentoring, you may find yourself once again in the role of student—or as teacher or mentor.

Mercury retrograde in Libra and the seventh house of marriage and partnerships. In this sign and house placement, communications issues may arise in a marriage or a partnership with a significant other—either personal or professional. It is a time to avoid making any kind of hasty engagements or wedding ceremonies, or forming an important business relationship. Rather, Mercury retrograde in Libra and the seventh house is more suitable for a "time-out" on major relationship decisions in favor of reflection and introspection. During this time period, for instance, you may want to give yourself permission to review your feelings for another person, addressing any doubts or anxieties. In an already-existing marriage or partnership, however, it is an ideal time for couples therapy, or for taking time away together—or apart—to process unconscious issues around the relationship. For in its essence, both Libra and the seventh house center on examining your self in relationship—where you give too much, how you hold back affection, or how you go about striking a loving balance with an intimate other.

Mercury retrograde in Scorpio and the eighth house of sexual intimacy, psychology, and occult study. Mercury retrograde in this placement can cover a spectrum of issues. At its core, this is the house and sign that rules the hidden side of life: sexuality, death and dying, the unconscious, or occult studies. Mercury retrograde here will want to draw aside the veil that separates the visible from the invisible, penetrating all the dark corners of the psyche—and the psychic realms. Thus it is a good time to pay attention to dreams or subconscious desires; to visit a medium or astrologer; or to explore your attitudes toward sexuality (traditionally, however, Mercury retrograde is not considered an ideal time to initiate a sexual relationship). Because Scorpio and the eighth house rules "other people's money," there may be issues around joint property or inheritance. As always, it is wise to observe Mercury retrograde's axiom, and take a wait-and-see attitude before signing any papers or making any joint investments.

Mercury retrograde in Sagittarius and the ninth house of religion, philosophy, higher education, and foreign travel. When Mercury retrograde falls in this placement, your focus may turn to the true meaning of life. Whatever you have based your truth upon becomes subject to review. As you reevaluate your philosophy in life, you may decide to take a college course, bury your nose in books on religion, or attend a conference on spiritual themes. You may decide to travel to a foreign country in order to study another culture (although it would be wise to keep in mind the possibility of travel delays or unexpected changes in your itinerary). With Mercury retrograde activating this sign and section of the chart, your soul may be possessed by a nameless and wordless restlessness. You may seek out a spiritual teacher who offers you the wisdom you are looking for. Indeed, it is an ideal time to deepen your spiritual practice, whether through prayer, meditation, or yoga. Likewise, you may wish to schedule a contemplative retreat in order to give yourself the time and space to delve into the remote reaches of the spirit.

Mercury retrograde in Capricorn and the tenth house of career and professional standing in the world. As Mercury retrograde retraces its steps in the sign and house of your professional field and public standing, you may undergo a period of revisioning. You may question whether your job is aligned with your higher calling or your purpose in life. Because of the increased attention to this segment of the chart, there may be career setbacks or misunderstandings on the job that require your attention. Rather than interpret these setbacks as negative, however, they can be seen as a kind of astrological "redirecting" toward your true work in life. On the other hand, as this is the most visible point in the chart, it is possible that you may receive some kind of public attention—either in a negative form like scandal or wrongdoing, or a more positive form of awards or recognition for accomplishments or work well done in the past. Once when Mercury retrograde was transiting my tenth house, a column I had written appeared as the lead story on the front page of a newspaper—a career first for me. Whether favorable or unfavorable occurrences take place under this transit; however, it is an important time to turn your thoughts and concentration toward your chosen work in life. Your work will be a source of learning, and you may even choose to go back to school, or take further training in your specific field of expertise. Knowledge of your specialty can be honed through work and study, as you perfect your professional capabilities.

Mercury retrograde in Aquarius and the eleventh house of friends and social organizations. Mercury retrograde transiting through this sign and house indicates that this is an opportune time to be with friends. You may feel compelled to strengthen your social ties, and reweave connections that may have become frayed due to the demands of work or family. Old friends may suddenly call or come to town for a visit. Your thoughts may turn to the subject of friendship in general. You may discover that you either lack friends, or have so many acquaintances that you haven't the time to cultivate connections of greater depth with a cherished few. In addition, it is a good time to review your membership in social or political organizations; you may ponder whether your needs for collective community are being met in a way that satisfies and that reflects your ideals. There may be parties, committee meetings, or planning boards to attend. On a deeper level, Mercury retrograde through Uranus and the eleventh house is about those dreams you hold for the future—both personal and collective. Thus it will catalyze a period of soul-searching around your hopes and ideals, and whether you have like-minded friends who share and support your vision.

Mercury retrograde in Pisces and the twelfth house of final endings and cosmic consciousness. Mercury retrograde in this house and sign may signify the ending of a significant life stage. For ultimately, Pisces and the twelfth house deal with issues around closure. You may feel as if you have come to the end of a job, career, or marriage. If you are young, you may be mourning the joyous days of youth, while if you are middle-aged you may be struggling with feelings of sadness around entering old age. Here Mercury functions as the mythic *psychopomp,* the guide who travels with us as we make the transition between worlds. As Pisces and the twelfth house symbolize cosmic consciousness, you may feel the need to be alone, and crave long periods of silence and contemplation. Your thoughts may be so deep, you may not even be able to put them into words; they may have a dreamlike cast and seem otherworldly, imaginal, and fanciful. As Pisces and the twelfth house are deeply sensitive, psychological, and introspective, this is a time to plunge into the inner world of the psyche and swim with your unconscious dreams and fantasies. It is a time to give full expression to the intuitive, mystical, and devotional side of your nature. As Mercury goes direct and you emerge from this ocean of consciousness, you will feel reborn, refreshed in body and soul.

Until this point, the preceding chapters have focused on laying the foundation for a more complete understanding of the mythological, astrological, and astronomical aspects of the planet Mercury and its retrograde cycles. The following chapters in **Part Two: The Path** offer a template for working with the cycle of Mercury retrograde from a more philosophical and spiritual perspective. Drawing on modern-day stories of Mercury retrograde, classical legends, literature, and symbols from the world's myths, this section will show the reader how to use this period of time as a way to enliven her creative imagination and deepen her connection to the inner life. Fully invoking Hermes-Mercury as our guide on a journey into the unknown, this section begins with an exploration of the stormy weather conditions that often launch the initial phase of Mercury retrograde, then moves into the deeper waters of the soul and creative spirit, and finally emerges once again into the "real world" with a new message and a fresh orientation to life.

PART TWO

The Path

6.

ON THE THRESHOLD OF
TWO WORLDS

And there shall be a great confusion about things,
what those things are, and where all those little
things lieth.

—Life of Brian

In ten minutes, exactly and approximately.

—Monsoon Wedding

Sing to me of the man, Muse, the man of twists and
turns driven time and again off course . . .

—The Odyssey, translation by Robert Fagles

Two worlds collide, two realities clash, when Mercury begins its retrograde phase. The drive to move forward and get things done in the usual fashion runs headlong into an on-coming force from the opposite direction that says, "Slow down— what's the rush?" Like the event organizer in the Indian film *Monsoon Wedding* who reassures the frantic father of the bride that he will be arriving to fix the wedding decorations in "ten minutes, exactly and approximately," schedules become quixotic and elastic. Nothing works the way it should, and misunderstandings abound. In short, we are no longer the capable masters of our universe. We enter a mythic time zone, where the familiar world intersects with an often turbulent and strange dimension of reality.

Indeed, when Mercury enters its retrograde cycle, we may find ourselves outside the boundaries of linear clock-time. As if we have been stopped in action in the freeze-frame of a film, we are being given an opportunity to pause and review the experiences we have encountered thus far along the road of life. For as Mercury retrograde cycles with their delays and interruptions remind us, the goals we seek are rarely quickly attained, but are more likely years in the unfolding. When Odysseus sets sail in *The Odyssey,* for example, he imagines that his journey home to Ithaca and his long-suffering wife, Penelope, will take him ten days. Instead, it takes him ten years. In the translation of Homer's epic by Robert Fagles, we are introduced to the homesick warrior-king with these poignant opening lines: "Sing to me of the man, Muse, the man of twists and turns driven time and again off course . . ."

This is a line so descriptive of life-as-it-really-is that it speaks directly to the soul in all of us. Who has not felt the frustration of being waylaid by impossible obstacles, or hijacked by forces outside of our control? When Odysseus sets sail for home, writes Bernard Knox in his introduction to Fagles's translation of *The Odyssey,* he is literally blown off the map of the real world "into a world of wonders and terrors, of giants and witches, goddesses and cannibals, of dangers and temptations." Seduced and bewitched by Calypso and Circe, shipwrecked by demonic gales, confronted by one-eyed giants, and driven deep into the underworld where he reclaims his will to live after being driven to despair and thoughts of suicide, Odysseus loses his way time and again.

Odysseus's journey has captured imaginations for nearly three thousand years because it so poignantly reveals the human condition in all its bewildering, confused glory of mystery and ambiguity. It is a journey we relive whenever we feel as if we have lost our way, or have forgotten our reason for living, thrown off course by winds of fate more powerful than ourselves. Running late, missing appointments, not making deadlines, forgetful of anniversaries and birthdays, chronically inconsistent and easily tempted—we are all like Odysseus, off course and late for life yet still, somehow, sailing toward home.

As uncomfortable as it may be for most of us, the days and weeks of Mercury retrograde are a time when the mythic reality of Odysseus intersects with the everyday reality of the ordinary person.

There is a word for this borderland state: *liminality*. The term comes from the Latin *limen,* and means threshold, indicating the transitional space in between two points. Thus when Mercury begins to slow down and move backward against the sky, it is a sign that we have entered an altered timescape where things do not work as they normally do.

If you have ever prepared for a long trip, you may have experienced a similar suspension of reality. As you fold your clothes, make last-minute phone calls, and firm up hotel reservations and car rentals, you know what it is like to dance in the space between worlds. Perched on the shore of your secure and known environment, you have not yet departed for the new city or country that is your destination. Expectantly looking forward to your travels, anticipating what experiences might lie in wait, and perhaps clinging a bit nostalgically to your old but familiar routines, you feel adrift on invisible currents. Liminality is also a phenomenon that occurs whenever we make the transition from one stage in life to another, whether packing up to move to a new house or city, marrying, having children, divorcing, or changing jobs.

Mercury retrograde may not coincide with a major life transition like marriage or a change in profession. Yet as things start to back up or we run into unforeseen obstacles, this murky interlude of liminality between "no longer and not yet" is just where we may find ourselves. For most modern-day people this limbolike state provokes unease. Accustomed to the security of Palm Pilots and DayTimers, a kind of vertigo can set in as we realize that we are not as in control of our lives as we would like to think. Caught in a time warp, frantic to return to our old normality, tension arises. Even those who know they are entering the period of Mercury retrograde can find themselves furiously competing against the clock to finish tasks, settle business matters, or resolve nagging relationship issues. Suddenly, the least important things assume significance out of all proportion, and a strange urgency fills the air. The atmosphere feels oddly unsettled, as if life is hanging in the balance. As words come out of our mouths in a way that we have never intended, confusion muddles our communications, and obstacles impede our progress.

Under such circumstances it becomes difficult to know how to manage practical affairs such as launching new business ventures, buying or selling stocks or other properties, planning for the future, making an emotional commitment, traveling—or simply conducting the day-to-day business of life. Astrology is not a dogma or a rigid theology that asserts hard-and-fast rules for how to handle planetary influences. As mirrored in the natal birth chart, the significance of each individual's unique situation far outweighs generalized theories. Yet most experienced astrologers offer their clients guidelines on how to manage mundane matters under the influence of Mercury retrograde. Like sailors tacking with the wind, they show us how to ride the currents of this turbulent time. As the following sections explore, what matters most as we enter Mercury retrograde is to downshift from our usual fast-track ways of thinking, regain our sense of humor, and accept with grace that our affairs may not go as smoothly as we would prefer.

SIGNING CONTRACTS, BUYING AND SELLING, TRAVELING, MAKING PLANS, AND MAKING LOVE

"When Mercury is retrograde," says Vedic astrologer Chakrapani Ul-lal, "the mind speeds up and creates a state whereby one's thoughts and actions race ahead of one's reality, thus leading to various forms of chaos or carelessness. That's why it is a time that astrologers advise against making important decisions, because important steps are often missed and details skipped over." When it comes to handling the practical side of life while Mercury is retrograde, most astrologers are generally in consensus with Ullal. "Just slow down and think about it for a while" is their oft-repeated mantra. Erring on the side of caution, they advise their clients against making major lifestyle decisions or purchasing mechanical or technological items such as a car or computer.

Astrologer Erin Sullivan, for instance, counsels clients that it is best to put off signing any important contracts or business agreements and to maintain patience for the entire three-week period. In *Mercury Retrograde: Your Survival Guide to Astrology's Most Precarious Time of the Year!* astrologer Chrissie Blaze imparts a similar message, warning that "if ever disaster could befall from a communication breakdown, this is the time. Don't sign up for cell-phone service, health-club memberships, car leases, or insurance policies . . . Don't even think about buying a house under MR (Mercury retrograde)."

Edith Hathaway, an astrologer who practices both Western and Vedic astrology, tells the story of clients who were supposed to close a deal on a house in December of 2002, just before Mercury was to turn retrograde. "All the papers were set," she recalls, "and the buyer had finally been approved for a mortgage." Thinking that everything was in order, Hathaway left on a trip for India. While she was away, she says, the sellers decided to postpone the closing until January 2003 in order to gain a tax advantage. Upon returning from her trip and learning of the new situation, says Hathaway, she advised her clients of the probability that there would be further delays and complications that would last at least until Mercury turned direct on January 22 (this retrograde period lasted from January 2, 2003, until January 22, 2003). "Sure enough," recalls Hathaway, "at the last minute it seemed that a subdivision requirement that the buyers had been assured by local authorities would be waived in this case, mysteriously could *not* be waived. All parties had to wait until the end of January for the local Planning Board to meet and approve of the subdivision."

Businesses such as auto and computer repair operations, however, says Hathaway, may actually flourish during a Mercury retrograde cycle—precisely because this is the time during which repairs on cars, computers, and communications equipment are more likely to be needed. Still, Hathaway cautions against launching marketing campaigns or doing big business deals, as they are subject to communications snafus like everything else. In addition, mistakes can occur on repairs that are underway.

Indeed, says astrologer Lynn Koiner, "never have anything repaired when Mercury is retrograde," as this cycle can have a devastating effect on anything related to transportation and communication. This has proved true in my own experience, when a seemingly simple repair can suddenly turn into something more extensive and expensive to fix. Twice I have taken my car in to have the oil changed while Mercury is retrograde; each time, the attendant forgot to replace the oil cap, leading to further repairs. Koiner also cautions against purchasing anything associated with transportation—be it a car, bicycle, computer, tape recorder, or even shoes.

Although I eschew any kind of wholesale advice, whenever possible I try to follow these guidelines. Given that our culture is so stressed and driven, it makes sense to go against the grain and carve out some time for thoughtful consideration of our actions. When asked to sign a contract with a newspaper service as a columnist during a Mercury retrograde phase, for instance, I confided in my editor the reason for my hesitancy on signing right away. Surprisingly receptive to my unorthodox wish, she agreed, and we waited together for Mercury to go direct before finalizing our agreement.

One reason astrologers urge their clients to exercise restraint during this time is that information is often withheld. This may not be because of malicious intent but the result of inadvertent forgetfulness or absentmindedness—two traits associated with Mercury retrograde. Thus when working on projects or making plans, it is wise to keep in mind that all the facts may not be available. Another reason to forestall major commitments, says Lynn Koiner, is that either one or another person involved may change his mind. Typically, as Mercury begins to station direct, information surfaces that had been considered inconsequential, "lost" or "forgotten," or people suddenly switch from one point of view to another. Thus any decisions, says astrologer Koiner, should only be made on contingency, as everyday commerce and business during this time is subject to "delays, last-minute postponements, or mix-ups."

More often than not, however, those decisions that cannot be put off will end up being delayed anyway—a natural result of the retrograde influence. In his section on Mercury retrograde in *The Inner Planets,* astrologer Howard Sasportas writes about an operation his surgeon had scheduled for him during a retrograde period. Due to his particular health-care arrangements Sasportas decided, after a great deal of worry and anxiety, that he had little choice but to go forward with the procedure. After checking into the hospital the day before the scheduled operation, writes Sasportas, he immediately ran into delays. When the surgeon showed up, he turned out to be a replacement for his regular doctor—who was late returning from a skiing holiday. That evening, an apologetic nurse informed Sasportas that his medical records and X rays had been misplaced—and that the procedure would have to be postponed. Departing happily, Sasportas "said a prayer of thanks to Hermes for hiding my files, which, by the way, were immediately located . . . two days later."

When it comes to making health-care and medical decisions during Mercury retrograde, Lynn Koiner advises those who are seeking a medical diagnosis or visiting a doctor for the first time to wait until the planet is direct. Often, says Koiner, she has observed many situations where an individual "goes from doctor to doctor seeking a diagnosis while Mercury is retrograde—only receiving a proper diagnosis when it turns direct." Those who may be forced to have surgery under emergency situations while Mercury is retrograde, however, says Koiner, should nonetheless be reassured that they can be treated successfully—as she herself has been.

Indeed, some decisions just cannot be delayed. Typically, against all dire Mercury-retrograde predictions to the contrary, things usually work out for the best—if not exactly the way we predicted. This was the case following my father's death. Due to many complex reasons, his estate had to be settled immediately; as fortune or the gods would have it, this coincided with Mercury's reversal. As executor of my father's will, I found myself having to take care of business calculated to set off any astrologer's alarm bells, including my own: going to court and signing legal documents, selling his house and car, liquidating his stocks—all major taboos from an astrologer's point of view. As an astrologer myself, however, I had the advantage of being able to consult my own birth horoscope to see what area of my individual chart the planet Mercury was currently influencing.

As it turned out, Mercury was moving through my twelfth house of spiritual consciousness and issues related to karmic connections and end-of-life matters. I took this as a sign that Mercury was literally leading me on a journey through the underworld, helping my father to conclude the remaining business of his life—and make the final transition to the next stage on his soul's journey. Everything went remarkably smoothly, and the affairs of his estate were settled quickly. As sometimes happens during retrograde phases, the boundaries between this world and the next were blurred, and signs, dreams, and portents proliferated. Browsing through the inspirational literature while waiting at the funeral home to make arrangements for my father's funeral, I casually flipped through a brochure—and was jolted to happen upon a quote from an article I had written some years earlier on near-death experiences. In yet another incident a potential customer showed up to buy my father's Lincoln Continental—with my father's *exact* name inscribed on his jacket, Joe Carroll.

Astrologer Dana Gerhardt also takes a more imaginative, open-ended approach to the transits of Mercury retrograde. For many years, she worked in a market-research firm that, she says, was very much "Mercury ruled"—signing contracts, organizing research projects, sending them out, and getting them back all the time. Gerhardt says that she found that few of the tried-and-true rules around Mercury retrograde applied, observing that "things went wrong when Mercury was retrograde—and things went wrong when Mercury was not retrograde." Similarly many things went right during that three-week period, she says, just as they did at other times. Employees hired under a retrograde phase, for instance, were just as likely to work out as those hired when Mercury was direct. However, says Gerhardt, while she doesn't believe problems are "ordained" during Mercury retrograde, she does like "entertaining the possibility that there will be problems with anything performed at that time—that way I'm in a better frame of mine to respond when the problem appears." My own experience as a journalist parallels Gerhardt's, and, when it was unavoidable, I often signed magazine contracts, and wrote and published articles during a retrograde phase. Just like Gerhardt, though, I took greater care with my actions, and put more thoughtfulness into my work.

Matters of the heart—whether a newly kindled romance, an ongoing relationship, or even family ties with children and siblings—can also take a strange detour while Mercury is retrograde. Over the years, I have observed that emotionally charged issues tend to surface during the days when Mercury is stationing retrograde or direct. Often when my sons were young, I would notice that arguments or disagreements over homework, curfews, or other misunderstandings tended to flare up at these times. Once, several weeks into a heady new relationship, the man I had been dating abruptly distanced himself from me just as Mercury stationed retrograde. Confused by his sudden about-face but wise to the ways of Mercury retrograde, I patiently waited—and waited. When finally my new love interest did call, his words were polite and cool as he mentioned the busy work schedule that prevented him from seeing me. Finally, as Mercury began to station direct, we went out again, and—as so often happens at the end of Mercury retrograde—the truth finally came to light as he confided his lingering interest in a previous relationship that hadn't completely ended, as well as his fears around commitment. After an in-depth discussion, and with the air finally cleared, we eventually went on to have a long-term relationship.

Tackling thorny intimate issues while Mercury is retrograde can be a two-edged sword: While it is a cycle that favors in-depth discussion, words can be easily misconstrued, we may blurt something out that surprises even ourselves, and the pot of emotional confusion we may find ourselves in may only become thicker. Thus I have found that it is best not to expect any kind of clarity or final resolution during this period. During Mercury retrograde, silence is indeed golden—especially when it comes to personal disagreements, it is sometimes best to let our problems simmer below ground while our psyche cooks up a solution.

Conversations and intimate dialogues within an already existing relationship are one thing. But many are curious about whether or not to begin a sexual relationship under Mercury retrograde. According to Gerhardt, launching sexual intimacy during this time may not be the wisest course of action, as that "puts Mercury retrograde in the event chart for the beginning of the relationship." In other words, Mercury retrograde will cast its influence over the relationship, and that, says Gerhardt, may "seed miscommunications" between the two people involved. On the other hand, says Gerhardt, a relationship that does begin under Mercury retrograde may be especially fated to teach or heal any issues the two partners may have around intimacy and communication. In general, however, advises Gerhardt, "break-ups or hook-ups during the retrograde cycle will very likely have an air of instability around them, and may unravel after Mercury goes direct." Thus, she says, the best advice would be to sit still without clutching too tightly to those we love, as, during the retrograde phase, "it's best to hold what you cherish with an open hand. You may get to see it in a new way."

As Gerhardt indicates, another effect of Mercury retrograde on relationships is that it may alter—for better or for worse—how we see our partner. We may see a side of our significant other that we had never glimpsed before: We may notice our partner's sense of humor, as if seeing it for the first time; or, conversely, we may notice what a biting sense of criticism he or she may have. In her book, Chrissie Blaze gives a positive example of a Mercury retrograde relationship that underwent such a process, telling the story of a happily married man who, on a whim, reconnected with an old flame on the Internet during Mercury retrograde. Seeing her picture, writes Blaze, triggered a flood of memories, and he thought of her obsessively. Rather than growing jealous, however, his wife took this as an opportunity to appreciate the old flame for how she had drawn her husband out emotionally and creatively. Grateful to his past love for what she had taught her man, writes Blaze, the wife realized that her husband's temporary regression into the past ultimately brought the married couple even closer together. Thus, when it comes to relationships, a wise use of Mercury retrograde might be to see it as a time to recognize the fragility and uncertainty that undergirds all our intimate kinship ties—valuing the mystery that brings two people together, for however long or short a period of time their destiny intends.

Travel—Mercury's favorite domain—is one of the trickier subjects to handle during a retrograde period. If possible, it's best to time departures and arrivals so that they do not coincide with either the stationary retrograde or stationary direct periods in order to avoid unexpected or lengthy delays. Otherwise, traveling under the influence of the planetary god of the journey should be pleasurable—as long as one is open to a change in itinerary. This holds especially true when it comes to making travel arrangements. One close astrologer friend, for example, hatched plans during a Mercury retrograde cycle with her partner to get away from their combined family of four children for a romantic rendezvous in New York City. They had a good excuse: the eightieth birthday celebration of her partner's mentor. But as quickly as the plans were made, they fell apart. First, the tickets she had purchased to hear the Dalai Lama speak were canceled due to his poor health. Next, while her partner was traveling to New York ahead of her, his mentor's health took a turn for the worse and the birthday party was canceled. Although they had dreamed of rejuvenating their passion while in New York, my friend reconsidered the whole idea and decided instead to take a trip to California by herself. Mercury in his role as the trickster, she concluded in an e-mail to me, "definitely had his way with us!"

MUCH ADO ABOUT NOTHING

If there is a moral to the stories of befuddlements, bafflements, and madcap mishaps that so often accompany Mercury retrograde, it is that most of what goes wrong falls under the category of the nonessential "small stuff" that gets us all down. No amount of mindfulness, it seems, can prevent Mercury from stirring up petty annoyances. Even while working on this section under a Mercury retrograde—taking my own advice and cultivating as much attention to detail as possible—I had mailed two separate letters, including money, to my sons. I had checked their addresses twice, as they had each just moved. Still, both envelopes were returned on the same day: one because it was a slightly oversized card that required extra postage, and the other because my youngest son had not given me his correct apartment number. In the larger scheme of life, this was a minor disturbance. Yet ironically, it was one of those small things that proved irritating and time-consuming—even as I was working to meet the deadline on a book on Mercury retrograde!

Many astrologers, in fact, ascribe the phrase "much ado about nothing" to the disturbances caused by Mercury retrograde. After all, cars break down, mail gets lost, and computers crash at all times of the year. Yet somehow, the confusion loosed during Mercury retrograde is doubly obtuse. The dysfunction and disarray seem to *matter* more during Mercury retrograde. Frustrations are greater, impatience is more pronounced, and the littlest things become the biggest things, the proverbial straw that breaks the camel's back.

One of my vintage Mercury retrograde experiences, for example, involved my vacuum cleaner. By nature, I am a person who thrives on the ritual of my daily routine. I read the paper, write, meditate, walk my dogs, and clean my house on a rhythmic schedule that rarely varies. Should one thing from my schedule go "off," my whole routine is thrown off and I feel unsettled. Several days into a retrograde cycle, the plug on my vacuum cleaner broke; appalled at the thought of the absence of my vacuum cleaner from my well-oiled life, I immediately drove it down to the local repair shop. Run by an eccentric, elderly lady, this repair center is a genuine curiosity shop, with decoupage gift boxes, ceramic pots, umbrella stands, and every model of vacuum cleaner imaginable stuffed into the corners. The pace is leisurely, as old-time customers chat up the proprietor.

The delightful "curiosity shop" atmosphere, however, was lost on me, as I fretted and fumed. After examining my plug, the owner filled out a receipt, saying it would be fixed by noon the next day. Promptly at noon the following day, I called, and was told that my vacuum would be ready by early afternoon, when I could pick it up. Preoccupied with work, I decided to delay picking up the vacuum. As I parked outside the store the next day, I immediately noticed that there were no lights on, even though it was the middle of a weekday. Puzzled, I got out and approached the door, where a piece of paper had been taped inside the plate-glass window. On it were hastily inscribed the words, "Closed until January 31st for much-needed R & R." I couldn't believe it—I had to live without my vacuum for almost the entire Mercury retrograde cycle! And how could a business owner just leave a note like that? In an instant, a seemingly small thing snowballed in my mind into an enormous inconvenience.

Fully aware that it was Mercury retrograde, however, I decided to lean into the wind, rather than go against the flow of energy. The message that had been left for me was clear: Take a personal time-out from cleaning. The first thing I did was relax my usual housecleaning rituals; in addition, I discovered the simple delight of the old-fashioned broom. Without my vacuum cleaner to nag me as a disapproving reminder, I discovered that I was less concerned with dust on the floors and took more time out to go walking every day. More than that, though, I realized I had been in a wound-up frame of mind, unable to let go of a large project I had just finished and manically trying to think up more things to be done. Now, with a hole torn in my all-important schedule, and thrust into the off-the-wall world of a vacuum-cleaner-repair shop, the larger message from Mercury seemed to be that I should relax my expectations and reconnect with my creative instincts.

Two weeks later, on the appointed date, I showed up at the shop, along with some other frustrated customers. Still, the store was dark and there was no sign of the owner. Finally, after an interminable eight more days of waiting—on the very day when Mercury was stationing direct—I returned to find the store reopened. My patience was tested once again as, while I stood in line for nearly an hour, the owner casually regaled customers with tales of her spontaneous, last-minute trip to visit her daughter. Practicing a kind of mental tai chi, I moved with the moment, rather than against it, and in the end delighted in the lazy afternoon conversations with complete strangers. As I waited in line, I examined the odd assortment of objects for sale stuffed in the store's nooks and crannies. My imagination was piqued by being in a business that was so obviously a fish-out-of-the-corporate-mainstream. When it was finally my turn, I was told that my vacuum *still* wasn't fixed. With the mantra "It's only Mercury retrograde, it's only Mercury retrograde" on my lips, I returned home, went back the next day, picked up my vacuum cleaner, came home, plugged it in—and watched it fall out of the socket. She had replaced the old plug with one that was too large! At this point, I could only laugh at Mercury's trickery.

Indeed Mercury, writes astrologer Robert Hand in his classic, *Planets in Transit,* is "the planet of pranks." His characteristically playful spirit is even further heightened during the planet's retrograde motions. Writer Narelle Bouthillier encountered Mercury's mischievous side while moving to a new apartment in Cambridge, Massachusetts. As she tells the story, she had made plans weeks in advance to pick up a fourteen-foot truck from U-Haul on May 31—just days before Mercury turned retrograde on June 3rd, 2001. With her current lease up on May 31, Narelle was scheduled to move into her new apartment on June 5th, as the woman occupying the room she was moving into could not move out earlier. Narelle had arranged to stay the few intervening days with a friend, while boarding her cats—at great expense.

When May 31 finally rolled around, says Narelle, she called U-Haul at 8 A.M. to see if she could go and pick up her truck, as her movers would be arriving at 2 P.M. As she recounts her telephone conversation at the time, "A man named Rich told me he had no trucks. I told him I had a reservation. His response was, 'So do you and fifty other people.' (The person with whom I made the reservation was no longer working there.) He said he was expecting trucks from Maine and that it would take six hours for those trucks to arrive. When I asked if they would arrive by 2 P.M., he said yes and that I should call back at that time."

When Narelle called back at the appointed time, however, the drivers still had not arrived and she was told to call back at 4 P.M. The hour came and went, and still there was no truck. By this time, says Narelle, she was so enraged, she was icily calm. At extra cost, she had no choice but to have the moving company take her belongings and place them in storage. Numb and stupefied, with no home and dwindling financial resources, she moved in with a friend for a few days until a truck became available. Finally, on June 5th, Narelle and her cats made the two-and-a-half hour drive to Cambridge, where she unloaded her belongings into her new apartment. Just when she thought the worst of her moving experience was over, however, Narelle's check to the moving company bounced. Thoroughly embarrassed and ashamed, and never having bounced a check before, she persuaded a friend to cover for her with his credit card. Several weeks later, Narelle inadvertently locked herself out of her new apartment, and ended up in the public rest room in a restaurant. Staring at herself in the mirror, she began to laugh, doubling over in mirth as she wondered what could possibly go wrong next.

All too often during Mercury retrograde, in fact, the joke really is on us. Sometimes the gods and goddesses *are* laughing at us—at our self-importance and ridiculous attachments to the most mundane things. If we are wise enough to drop our defenses and get the joke, we can chuckle alongside the gods and enjoy the lightheartedness that can arise from not taking ourselves—and our lives—so seriously.

TRICKSTER AS SPIRITUAL TEACHER

In his tendency to whip up a tempest in a teapot, we see Mercury in his wily trickster aspect. For while the disturbances we experience during Mercury retrograde may center on life's small stuff, they contain big spiritual lessons. The pranks Mercury plays on us are not merely hollow tricks but teaching devices through which we can become more wise to the ways of the human condition.

The trickster figure, in fact, occupies an important role in myth and religion. Whether in the role of court jester, holy fool, or clown, the trickster's function is to puncture the human tendency to become inflated with one's own importance. The impulse to be impish in the face of moral priggishness, to overturn religious traditions when they have become too stuffy and intellectually remote from the human condition, is the function of the trickster archetype. Acknowledging the need for playful irreverence as a necessary counterpoint to religious self-righteousness, many traditions have made a special place for the trickster as teacher of spiritual truths.

At a particular stage during the sacred dances of the Native American Maidu of Northern California, for instance, *Coyote* appears dressed in black feathers with sticks for legs and proceeds to poke fun at the somber proceedings. In his essay *On the Psychology of the Trickster-Figure,* Carl Jung describes medieval church celebrations called Fool's Feasts, or Fool's Holidays, when the religious hierarchy was reversed. Children were dressed as bishops and priests; in some instances masqueraders disguised as animals and women interrupted church services with games of dice, burned foul-smelling incense made of shoe leather, and ran around the church. In Sufism, the *Madzub,* the holy but naive fool crazy with the love of God, utters words more true and sane than the ordinary person full of self and blinded by worldly illusion. Zen Buddhist teachers are famous for their puzzling *koans,* or parables, that reveal the paradox and ambiguity at the heart of life. The well-known Zen aphorism "If you meet the Buddha in the road, kill him" is a spiritual teaching in trickster form. Its purpose is to point out that even the holiest of beliefs, when blindly adhered to, can become a hindrance on the path to further enlightenment.

Often, the trickster archetype is revealed in nature's capricious forces—the sudden storms, droughts, floods, and earthquakes that bedevil our existence. As a child growing up in the Midwest, I recall being terrified by what seemed to me the numinous, supernatural powers of the tornado. When the warning siren sounded, my family and I would collect our things and vanish into our underground cellar until it was safe, sometimes not knowing what would greet us upon opening the cellar door. The tornado's uncanny habit of touching down in a particular spot, pulverizing everything in its wake save the odd desk with an open book or a kitchen table set for a meal, was unnerving. With its power to destroy my secure family life, the tornado seemed like a very real and crafty being to me at the time, and I titled my first childhood story "The Tornado Devil."

My childhood fear that saw the devil's hand in the wild forces of nature reveals the human tendency to ascribe evil to those things that cross our path and block our way without warning. Indeed the original meaning of the Hebrew term *satan,* writes scholar Elaine Pagels in *The Origin of Satan,* is not the name of a particular character but a form of adversary actually sent by God to obstruct our activity. The root of the word *stn,* she writes, means "one who opposes, obstructs, or acts as adversary," while the Greek *diabolos* literally meant "one who throws something across one's path." Thus the *satan's* presence in the Hebrew Scriptures provided a divine explanation for sudden reversals of fortune, allowing God to block a path of action that a human being could not see for himself was bad.

Pagels gives as an example the story of Balaam in the biblical book of Numbers, who had willfully decided to go where God had ordered him not to. Saddling his ass, Balaam goes anyway. While Balaam is ambling along the road, God sends an angel with a drawn sword to stand in his way—yet only the ass can see it. When the ass turns aside from the road in fear of the great being before it, Balaam strikes her to make her get back on the road. Once again, the angel appears, causing the ass to balk; yet again Balaam strikes his ass. The third time the dutiful but confused ass sees the obstructing angel, she stops and lies down in the road. This time Balaam angrily strikes her with his staff, leading God to put words in the ass's mouth, who says to Balaam, "What have I done to you, that you have struck me three times?" (22:28). After this, Balaam's eyes are opened by God, and, seeing the angel before him, he falls to his face. The *satan* speaks, saying, "Why have you struck your ass three times? Behold, I came here to oppose you, because your way is evil in my eyes; and the ass saw me . . . If she had not turned away from me, I would surely have killed you right then, and let her live" (22:31–33).

In her book, Pagels describes the Biblical *satan* as he appears in the book of Job as a kind of "roving intelligence agent" sent from the heavenly court, indicating that there is a universal intelligence wiser than ourselves, if only we would trust it. For just like Balaam, we often willfully resist the heavenly messenger in the road that thwarts our progress, angrily imagining a devil where a guardian angel really stands. Only the very wise among us, it seems, can glimpse the presence of divine protection in the limits that are sometimes placed upon us. Especially during Mercury retrograde, we may find ourselves in the same position as Balaam: cursing the obstacle in our path rather than blessing it as an angel of the Lord—the helpful *satan* preventing us from a course of action that just may not be right for us at a particular moment in time.

LIFE AS A JOURNEY

Especially during the first week to ten days of the retrograde cycle, when Mercury stirs the pot of our practical affairs, we may despair of ever reaching our goals. Faced with setbacks, we may feel as if we are running in quicksand. Whatever personal quest we are on may seem to recede into the distant background. One close friend, for instance, attended a writers' conference during Mercury retrograde. While there, she met a literary agent who offered to take a look at her book proposal. After reading it, the agent told her the idea was weak and that she should come up with something entirely new. So discouraged did my friend become that she nearly decided to give up writing altogether. After Mercury turned direct, however, she realized that the agent was simply not right for her. Regaining faith, she began working enthusiastically on her original idea—yet from a completely different approach.

Similarly, I have sometimes found myself stuck or blocked during Mercury retrograde; once, while working on a tight book deadline, a friend became seriously ill, and another friend showed up on my doorstep in the grip of a severe bipolar depression. The real-life personal crises of those I loved and cared about, of course, were far more important than a writing project. Still, I felt frustration, not knowing how I would handle these competing demands on my time and attention. Recalling *The Story of the Other Wise Man,* by Henry Van Dyke, however, helped me cultivate a different attitude toward the obstacles I encountered along the road of my life's journey.

Unlike Odysseus, who finally reaches home at the end of his saga, the hero of this tale ultimately never reaches the goal he so passionately seeks—at least, not in the way he imagines. Rather, he finds his dream along the way, in the inexplicable encounters that fall across his path. This classic little gem of a story with astrology woven into its narrative tells the tale of Artaban, a Zoroastrian Magus. Steeped in the knowledge of the stars, Artaban has a burning desire to meet the long-awaited Christ child. Artaban's odyssey begins when he receives word from his three peers in Babylon of the planetary conjunction that will soon take place, heralding the birth of a world savior. On the eve of this long-anticipated stellar phenomenon, they have invited him to journey with them to Jersualem to greet the new savior with rare gifts. In preparation, Artaban has sold all his belongings and purchased three jewels—a sapphire, a ruby, and a pearl—to offer as tribute to the newborn king. The three companions, however, can wait for him only ten days before leaving on their journey.

After beholding the sign in the sky, Artaban departs from his home quickly, riding steadily through the day and night, until he draws near to Babylon. On the outskirts of the city, however, Artaban is stopped in his tracks by a dying man lying across the road. Turning away from the man's cries, Artaban continues on his way, for he knows that if he lingers the Magi will leave without him. Pierced by the man's cries, however, Artaban prays for guidance, asking God whether he should risk, as Van Dyke writes, "the great reward of his divine faith for the sake of a single deed of human love?"

Still unsure of the right course of action, Artaban nonetheless dismounts and saves the man's life, offering him a special healing draft from his travel pouch. When he reaches the Temple of the Seven Spheres where he was to meet up with the caravan, he finds that the three wise men have already departed for Jerusalem, leaving him to cross the desert alone. Without their support, Artaban is forced to sell his sapphire to purchase his own train of camels. Setting off across the desert, he travels quickly until finally reaching Bethlehem, arriving at his destination just three days after the Magi, who have already offered their gifts of gold and frankincense and myrrh to the new king, Jesus. Hopeful at the prospect of finally catching up with his companions, Artaban stops for a rest at the home of a young mother with a newborn son. There, he discovers that the Magi, forewarned of an invasion of Roman soldiers, have fled the city along with the holy family. When the soldiers of Herod approach the home of the young mother where Artaban is staying, he blocks the entrance of the doorway, protecting the child inside from murder. He bribes the captain from carrying out his murderous mission with the second of his jewels, the ruby. Even the joyful sobs of the mother whose child he has saved, however, cannot prevent Artaban from wondering, writes Van Dyke, whether he has taken the right course or whether he has "spent for man that which was meant for God."

For the next thirty-three years, Artaban wanders the deserts of the East, a pilgrim in search of his king. Somehow, he never manages to catch up to the Magi or his Savior. An old man on the brink of death, Artaban arrives one last time in Jersualem during Passover. Swept up in an excited crowd, he is told that the famous teacher Jesus of Nazareth is to be crucified in Golgotha. Hope fills his heart, and the thought comes to Artaban that perhaps he can use his last remaining jewel, the pearl, to ransom Jesus' life. Following the crowd, Artaban is swept along toward the gate of the city when suddenly he witnesses a young girl being dragged by soldiers down the street. Suddenly, the girl breaks away and throws herself at his feet, begging him to save her. Recognizing the Zoroastrian symbol of the winged circle on Artaban's breast, she tells him that she is a daughter of the Magi, and is being sold as a slave to pay off her father's debts.

Faced once again with the "old conflict in his soul," writes Van Dyke, Artaban doesn't know whether what faces him is his greatest opportunity—or his last temptation. The one thing that *is* clear is that what has just happened is inevitable and, as he asks himself at the conclusion of the story, "Does not the inevitable come from God?" Taking the pearl, which he has kept close to his heart all these years, from his breast pocket, Artaban hands it to the slave as her ransom, delivering her into freedom.

As he does so, an earthquake rocks the earth, the sky darkens, and Artaban realizes that he has failed in his quest. At the same time, a peaceful feeling of resignation comes over his spirit as he realizes that he has done the best that he could from day to day. "He had been true to the light that had been given to him," writes Van Dyke. "He had looked for more. And if he had not found it, if a failure was all that came out of his life, doubtless that was the best possible." At that moment in the story, a heavy tile strikes Artaban on the temple, and he begins to die, slowly losing consciousness. Through the darkness, a light grows and a faint voice comes to him, saying, "Verily I say unto thee, Inasmuch as thou hast done it unto one of the least of these my brethren, thou hast done it unto me."

Though Artaban spent decade after endless decade trying to

catch up to the other wise men, his roundabout detours in response to the suffering of those who fell across his path brought him in the end to the place he sought to be. As Van Dyke concludes Artaban's story, "His journey was ended. His treasures were accepted. The Other Wise Man had found the King."

As the parable of the fourth wise man reveals, some of the most important journeys we take in life—whether a relationship, a creative project, or a job—are those where we become lost, disoriented, and sidetracked from the destination we are trying so valiantly to reach. In his book *Eros and Pathos: Shades of Love and Suffering,* Jungian analyst Aldo Carotenuto sheds light on this mystery. He recounts the story of the old soldier who was looking for his heart. When a wise man told him it was at the other end of the world, he went there—but didn't find it. When he returned, writes Carotenuto, the wise man told him that in fact he had found it by going on his journey.

DAYS DROPPED OUT OF TIME

Literature, with its long and storied view of life, offers a useful paradigm for handling the vicissitudes of Mercury retrograde. Like Odysseus, like Artaban, or like the old soldier, we may be blown off course or never reach our goals at all. But if we engage the odd encounters and strange byways down which we find ourselves wandering with imagination and courage, we discover that the treasure we seek is concealed within the journey itself. Then, like the ending of all good quest stories, we may find ourselves in a place richer with meaning than anything we could have dreamed of in the first place. As Cavafy's lyric poem phrases this truth, "Ithaca has given you the beautiful voyage. Without her you would have never set out on the road." For this reason, struggling against the incoming and outgoing tides of change is ill-advised under Mercury retrograde. Rather, like surfers who ride the waves, or sailors who steer their ships in the direction of the wind, it is an opportunity for practicing the spiritual art of letting go—and going wherever the wind blows us.

Reframing the impatience and frustration that arise over all the little annoying things in life that can and do go wrong is a powerful form of spiritual discipline. Acceptance for what is, rather than how our limited ego would like life to be, is a dress rehearsal for those inevitable periods of disruption in our lives when we are *really* faced with major life transitions. That is why spiritual practices are called "practices." As we practice acting out of an expanded state of consciousness instead of our usual narrow perspective, our souls are increasingly prepared for the real event—the eventual ego deaths, losses, or sudden successes when our worlds are turned upside down and inside out.

Indeed, we can look to Mercury retrograde as a kind of life *sesshin,* or Buddhist retreat, in which we sit silently on the banks of the rivers of our lives simply observing events as they flow downstream. Practicing "being Buddhas" we disengage from our attachment to things being the way we think they ought to be. "No appointment, no disappointment," advises Buddhist teacher Dean Sluyter in *The Zen Commandments.* "If you don't build your world on expectations, it doesn't collapse when things turn out different." This is not resignation, he continues, "but liberation. It's disillusionment in the best sense: cutting free from the illusions that have bound us. And since all our expectations are based on what has gone before, it's opening up to amazing possibilities, to that which we don't know how to expect."

It is in just this condition of "beginner's mind" that new insight and growth can occur. As biologist Edward O. Wilson explains the emerging scientific paradigm of compexity theory in *Consilience,* systems that are in perfect internal order, such as a crystal, are finished products that cannot evolve beyond their present state. A completely chaotic system such as boiling water, however, has too *little* structure to allow it to change. It is the system that exists on the "edge of chaos"—containing order, yet loosely enough structured to allow change to take place—that evolves most rapidly.

The potential contained in the "edge of chaos" is captured in one of my favorite novels from girlhood, *The Moonspinners,* by Mary Stewart. In the opening paragraphs the heroine arrives for a vacation in a small town in Greece a day ahead of her planned schedule, before her aunt is to join her. This leaves her with the delicious feeling, as Stewart writes, of "a day dropped out of time." That phrase, so redolent of a sense of magical expectancy, has never left my imagination. And in fact, it is during this sudden windfall of unplanned time that, in the novel, a mystery unfolds, changing the course of the heroine's life.

If we let them, the first week to ten days of Mercury retrograde can be like "days dropped out of time." As often happens in this initial phase, a twist of events may throw us off our schedules, things may fall apart, and we may find ourselves on the edge of chaos with the archetypal "edgeman," Hermes, as our guide. Embracing the unknown, we depart the one-dimensional flatland that our overscheduled lives can sometimes be, entering the realm of the uncanny and the unplanned.

7.

NOW LET US
BE SILENT

Now let us be silent
So that the Giver of Speech may speak.
Let us be silent
So we can hear Him calling us
    secretly in the night.

    —Rumi, *The Birds of Paradise,*
    translation by Jonathan Star

I have told you again and again—
    go to that inner silence!
    But still you do not hear me.

    —Rumi, *Don't Sleep,*
    translation by Jonathan Star

In the ancient Mediterranean world it was thought that whenever a pause in conversation occurred, the spirit of Hermes was nearby. A precursor of the belief that a break in the flow of talk signals an angel passing overhead, this time-honored custom says that silence gives voice to the divine presence. Perhaps the reason Mercury retrograde touches a chord among so many these days is because it awakens something old and long forgotten in our psyches—the need to periodically renew ourselves in the waters of solitude and reflection.

The initial disorientation that accompanies the first phase of Mercury retrograde launches us on a journey away from the distractions of the outer world inward toward the depths of soul. Any restlessness and unease we may feel at this time is a signal that we are being called to shift from mundane to sacred time. If we "hear" them right, the annoyances that may crop up are calls to set aside our preoccupation with the transitory things of this world and venture into the heart of silence. As we settle into the rhythm of this inner pilgrimage, we begin to take pleasure in the new spaces of quiet and calm that open up. We journey, writes Rumi, "from self to Self and find the mine of gold."

Perhaps more than at any other time of the year, the lure of solitude beckons during Mercury retrograde. The need to dive into one's innermost being, to embrace the mystery of aloneness at the core of life, looms greater. To move with this natural rhythm is to harmonize with the graceful dance of our own souls. Meditative silence is the foundation of the mystical life; without quiet we cannot hear the "Giver of Speech," as Rumi says. The twentieth-century Sufi mystic Hazrat Inayat Khan wrote often of the value of repose, regarding it as a great mystery of which most people are ignorant. Repose, he wrote in *The Inner Life,* "makes possible to learn by one day of silence what would otherwise take a year of study; if only one knows the real way of silence." If our lives allow it, Mercury retrograde lends itself well to cultivating hours, days, or weeks of thoughtful reflection. Attending a spiritual or religious retreat or spending time alone in nature are also ways to draw out the golden thread of silence woven into this cycle. If we cannot take the time to go away—and most of us can't—we *can* choose to make our inner life our priority, setting the needs of our psyche above the demands of the world. By giving extra care and attention to our meditation practice, yoga discipline, or prayer life, we can answer the call of silence arising from within. In this way, we work with the stars to reset the clock of our soul according to the timelessness of the universe.

For some, the notion of withdrawing attention from the duties of everyday life may seem irresponsible. One reason for Mercury retrograde's widespread reputation as an agent of confusion may be the profound resistance against slowness that is so deeply entrenched in our society. Thus our very defensiveness against inwardness and introspection can be the catalyst that can cause things to break down or go wrong. Everything about Mercury retrograde—ambiguity, disorientation, indifference, dreaminess, and the need for solitude—goes directly against the grain of the contemporary culture we live in. The modern world seems to recognize only one pace: busyness.

In his foreword to Stephan Rechtschaffen's *Timeshifting: Creating More Time to Enjoy Your Life,* Thomas Moore writes that in his readings of the literature written before the twentieth century he never came across the oft-repeated cliché one hears or reads so often today: "I'm just too busy." The "busy" fantasy, says Moore, seems to be part of modern life, though people in the past accomplished just as much without timesaving technologies. Indeed, throughout most of prerecorded history, writes social critic Jeremy Rifkin in *Time Wars,* our ancient ancestors calculated time according to the more elongated rhythms of nature, such as the migratory patterns of the animals, the ripening of the berries or roots they depended upon for sustenance, or the change of seasons. In the high-tech era, writes Rifkin, even the clock has become a relic of another, slower era as we now pace our lives to the rhythm of the computer "nanosecond."

For millennia, wise men and women have taught that true contentment lies in the simply lived life, of which unhurried, unrushed time is a major component. Yet despite the current fascination for yoga and meditation, this lesson still goes unlearned. The fact that stress is a leading cause of illness and that all our material wealth has not led to a corresponding rise in happiness is a revealing commentary on the kind of society we live in. The link between relentlessly overscheduled, fast-paced lifestyles and ill health led Stephan Rechtschaffen, M.D., author of *Timeshifting,* to diagnose a uniquely modern malady he labeled "hurry sickness." What was the treatment Dr. Rechstaffen prescribed for such a condition? Large doses of solitude and relaxation.

Increasingly, more and more people in the business world and other professions are recognizing the value of alternating rhythms of time. An article by Rebecca Bryant in the February 2000 issue of *Business 2.0,* "Two-Timing the Clock," advises busy professionals to turn back the clock by dividing temporal experience into "slow," "fast," and "middle" lanes. Bryant recommends minimizing intervals in the fast lane where much of pop culture and the business world operates, while maximizing time spent in the middle lane of family, church, and education, and the even slower lane of creativity, relationships, and spirituality. Consultants on integrating spirituality into the workplace have begun teaching overworked professionals the worth of incorporating a few moments of silence into their workday, whether through observing the ebb and flow of one's breath, taking a solitary walk during lunch, or silently repeating a familiar prayer or mantra at various times during the day.

WITHDRAW AND RETURN

When seeking the solace of silence and slow time—especially in the interlude offered by Mercury retrograde—religious traditions offer an abundance of wisdom. Traditionally, the world faiths have built in to their daily and yearly calendars rites and rituals that demarcate the boundary between the demands of the everyday world of work and family and the spiritual dimension that addresses the larger concerns of the soul. Whether answering the Muslim call to prayer five times a day, attending daily mass, meditating in the morning and evening, or going away on a spiritual retreat, these time-honored practices embody within them a healthy body/mind wisdom. By varying our rhythms, interweaving solitude with busyness, we balance the needs of our inner and outer lives.

Religious historian Huston Smith, for instance, wrote of the pattern of "withdraw and return" that is "basic to creativity in all history." In his discussion of Buddha in *The Religions of Man,* Smith described how Buddha withdrew for six years, then returned for forty-five. The "silent sage" similarly divided each year into periods of engagement and seclusion, spending nine months in the world, and the three months of the rainy season in retreat with his monks. Even Buddha's daily cycle, wrote Smith, was molded along the pattern of withdrawal and retreat, and "three times a day he withdrew that through meditation he might restore his center of gravity to its sacred inner pivot."

The Buddha's First Noble Truth that all life is suffering sheds light on the psychological mechanism that lies behind the need for periodic retreat and withdrawal. Suffering is translated from the word *dukkha,* which was a Pali term referring to an axle that had slipped off center from the wheel, or a bone that had been wrenched out of its socket. Thus suffering could be said to originate in that restlessness that arises from being disconnected from our center. "Life in the condition it has got itself into is dislocated," writes Smith in *The Religions of Man.* "Something has gone wrong. It has slipped out of joint. As its pivot is no longer true, its condition involves excessive friction (interpersonal conflict), impeded motion (blocked creativity), and pain."

The repeat pattern of Mercury retrograde throughout the year offers us a cosmically timed schedule to "withdraw and return" in order to realign ourselves with the spiritual center from which we have been dislocated. Astrologers often refer to Mercury retrograde periods as anything that begins with the prefix *re,* such as revise or review. But from a spiritual perspective the ultimate goal of Mercury retrograde is the encounter with *dukkha,* or suffering, and the return to the center of oneself. A song I used to sing during Sufi gatherings is a haunting lament of this need: "Return again, return again, return to the home of your soul. Return to who you were, return to where you were born . . ." Sitting with my fellow seekers in meditation, I would feel the sweetness of the longing that arose as, like a circle of sunflowers, we reoriented ourselves toward the luminous true north of our souls.

The center we seek, of course, is both within and yet everywhere at once. It is the circumference of the circle, and the dot in the middle. The sacred center exists in the present moment and outside time, in eternity. We are all, so often, every day, tested, distracted, and thrown off course far, far outside ourselves. The suffering caused by this separation is only healed when we undertake the journey back to the center, to the seat of our soul-truth and unity with the universe.

All meditation and sacred practice serve as *reminders* of our original place in the cosmos—the true "whence and whither" of life.

Most people have a sharply honed sense of just what this sacred, silent "center" feels like—if only they can be still long enough to tune in to its stabilizing wisdom. Yet paradoxically, it often takes a journey to realize the treasure in our own backyard, or the soul within the temple of our bodies. Something about a pilgrimage, whether inner or outer, it seems, deconstructs our habitual ways of thinking. Thus the Australian ritual of the walkabout or the Native American vision quest in which an individual sets out into the wilderness is a useful metaphor for a Mercury retrograde return to soul, summoning the god Hermes in his role as soul guide. Even if we cannot make a formal retreat under the guidance of a teacher, or spend our weekends in nature, we can loosen our ties to the world just enough that our attitude becomes one of a wanderer or a seeker. In a walkabout or a vision quest, for instance, the senses are opened to the universe, rather than exclusively focused on a narrow set of goals to achieve. To see ourselves on a journey lightens our mental burden of responsibilities. Synchronicities and surprises that we may have overlooked before— the chance encounter with a remarkable stranger, a saffron-and-rose sunset, or the sparkling joy of a child's smile—gleam out from the gray colors of mundane life. Indeed, though we may think that we have to break from our settled life in order to take a spiritual journey, writes Hazrat Inayat Khan in *The Inner Life,* "there is no one living a settled life here; all are unsettled, all are on their way."

By allowing Hermes to lead us in his role as soul guide, it is possible to go about our daily lives as if we were on a walkabout or vision quest. We do this by cultivating a receptivity to the inner dimension of the external world, thus finding the center within wherever we are and whomever we are with. Mystics of all ages have written of the divine as a numinous *presence* that is manifest within all things. Listening through the ears and seeing through the eyes of the soul, we heighten our awareness of that presence in our lives, allowing the invisible world to shine through the visible.

A lyrical Australian movie from the early seventies, *Walkabout,* illustrates this mood I am trying to convey. It tells the story of a young girl and boy abandoned in the Australian outback. Rescued by an aborigine, they spend a period of time in the pristine wilderness; though frightened at first, these urban children gradually open up to the stunning world of nature around them. Mostly silent, the movie unfolds in a series of images that picture their journey through the desert as a communion with night, waterfalls, wild animals, the burning sun, and the thick net of stars—a wondrous *revelation* of the miracle of existence. After their eventual return to civilization, a series of flashbacks on the part of the young girl as a grown woman shows how, despite the passage of years, she yet retained within her the memory of her sweet, solitary time alone in the wilderness.

As children of the universe, each of us carries within our beings a faint echo of the birth of creation; if we listen closely we can hear the strains of that deep sound. One of my favorite pieces of music to meditate to is called *Hearing Solar Winds,* by the Harmonic Choir. A tapestry of overlapping harmonic overtones, it is the music of cosmic consciousness that permeates all life. If you listen carefully as you go about the business of your day, you can catch this background noise of eternity, murmuring and humming like a distant ocean. "Enlightenment," wrote the Hindu visionary Krishnamurti in *This Light in Oneself,* "is not a fixed place. There is no fixed place. *All one has to do is understand the chaos, the disorder in which we live.* In the understanding of that there comes clarity, there comes certainty . . . That certainty is intelligence . . . It cannot be put into words because the word is not the thing, the description is not the described. All that one can do is to be totally attentive . . ."

TWO DROPS OF OIL

To return to the center within the infinite circumference of the soul while still going about our business is what Jesus referred to as "being in the world but not of it." Like tai chi or chi kung, this is a delicate dance of balance to achieve. There is a parable within the fictional parable of *The Alchemist,* by Paulo Coelho, however, that helps to show the way. In the book, it is handed down from the Old Testament prophet Melchizedek to the main character, a young shepherd who is about to set off in search of his destiny. Offering the youth a bit of wisdom to help guide him on his way, Melchizedek tells the eager pilgrim the story of a boy who is sent to learn the secret of happiness from the wisest man in the world. Rather than in a retreat hut, this saintly man lives within a castle that is a beehive of activity, with music, food, and tradesmen coming and going. Telling the young boy he is too busy to speak with him, the wise man suggests to the youth that he look around the palace. At the same time, he asks him to do him a favor and carry a teaspoon with two drops of oil without allowing it to spill. With his eyes focused on the spoon, the boy explores the palace, climbing and descending stairs. When he returns after two hours, the wise man asks him if he enjoyed the precious tapestries, gardens, and parchments throughout the palace. Too preoccupied with not spilling the oil, the boy shamefacedly confesses he saw nothing.

So the wise man sends the boy off again, admonishing him for missing "the marvels of the world." This time, the boy feasts his eyes upon the art, flowers, and archtitecture. Returning to the wise man after two hours, he realizes he has forgotten to keep still the spoon he was supposed to hold—and that the oil he was asked to keep is gone. " 'Well, there is only one piece of advice I can give you,' said the wisest of men. 'The secret of happiness is to see all the marvels of the world, and never to forget the drops of oil on the spoon.' "

If you have ever observed a spiritual teacher in the act of walking, you will get a feeling for what the concentration of carrying "two drops of oil" is like. The Vietnamese Buddhist monk Thich Nhat Hanh, for instance, leads "mindfulness" walks with his students. As he guides students through parks and along sidewalks, he carries about him an atmosphere of presence like a mantle of silence. He is both in the world, yet conscious of an unseen dimension at the same time. When leading his meditations, Pir Vilayat Inayat Khan often reminds his students of the Sufi dervishes whose consciousness was so immersed in the divine that they could be sitting on a garbage heap or confined in a prison, yet still maintain consciousness of their unity with the cosmos. Likewise, the Chishti Sufi teacher, Murshid S.A.M., a famous figure in the San Francisco Bay area during the late sixties, used to lead his students on spiritual walks of attunement right through the crowded pathways of Golden Gate Park. In the same way, walking as a spiritual practice is a metaphor for being in the midst of life while staying centered within the sanctuary of our souls. Thus one could say that the secret to a meaningful Mercury retrograde retreat is to continue to go about our lives—while still carefully tending the "two drops of oil" that are the silent kernel of our innermost selves.

MORAL REFLECTION

One of the pieces of advice astrologers give their clients during Mercury retrograde is the injunction to delay making important life decisions. This may not always be possible. But there is something sage in this counsel—the notion of thoroughly thinking things through before undertaking a significant change. Although the kind of silent, uncluttered time that lends itself to depth of thought is an increasingly rare phenomenon, Mercury retrograde is an optimal time to introduce the quality of thoughtfulness into our lives.

Often, for instance, a disagreement, an ill-timed remark, or a misunderstanding at the outset of Mercury retrograde can be the catalyst for soul-searching. Like a woman I know who broke off an affair she was having and decided to take a retreat in a monastery, we may be wrestling in our heart with issues around love and commitment. Or, like the man I know who was angered by his boss's seemingly unnecessary hard criticism of his job performance, we may be struggling with whether or not to quit a job or find a new place to work. But whatever the reason, Mercury retrograde is not the time to make impulsive decisions, or to shoot from the hip. Rather, it is a time to simply stop and think—to step back and mull things over, while increasing our capacity to be guided in our words and our actions by our inner voice. Building silent hours into our mornings, evenings, or weekends while the planet of communications is retrograde is a powerful way to work with the stars to deepen our thinking processes. Likewise, studying a timeless work of philosophy or metaphysics can broaden our range of thought, allowing us to reflect more creatively on the moral conflicts we may find ourselves grappling with. Brooding over our problems without the distraction of having to arrive at an immediate solution, as well, helps us in the end to arrive at a conclusion that is not impulsively reactive, but well thought out. Those insights that have been deliberated and pondered over time, and weighed and measured for depth and sincerity, are as precious as real jewels.

The link between moral integrity, depth of thought, and solitude has a long history. As legal scholar Stephen Carter writes in his book *Integrity,* Plato drew attention in his dialogues to the fact that Socrates would "quite literally *stop* to think—stop walking, stop his remarks." By this, writes Carter, "we are surely meant to understand that Socrates is more admirable for this habit, that what he says after a suitable pause is deeply thought out and just as deeply meant."

The image of Socrates pausing mid-sentence while suddenly immersed in meditative thought evokes the centrality of reflection to the moral and philosophical process. Great philosophers such as Ludwig Wittgenstein and Immanuel Kant, for instance, never married, and nurtured their capacity for original thought in long stretches of quietude and study. Silence, it seems, is an elixir that allows us to distill sense and order out of the unpredictability of our lives. For while we may turn to others for help in solving the problems of our lives, ultimately our most important decisions are reached alone, in the seclusion of our souls and guided by the inner light of spirit. Only think, for instance, of the poignant picture of Christ bent in prayer through the long night awaiting his betrayal and crucifixion while his disciples slept; the knight of Arthurian England kneeling in silence in the cathedral before undertaking his quest; Joan of Arc in rapt communication with her visions before advising the young French king, the dauphin; or Rodin's powerful image of human reflection, *The Thinker.*

Like the clear sound of a bell, something about the still pause or the silence of a solitary hour reflects purity of moral intention. If we pause to consider why this is so, it is not just because we think highly of those moral paragons who always seem to do the right thing. Rather, we admire most those individuals who may not always choose rightly, but who seem willing to devote time and energy to choosing the wisest course of action amid life's maze of confusion. The woman who agonizes over whether to get a divorce; the politician who thinks long and deep before deciding to commit the country to war: We may not agree with their ultimate choice, but we respect them for not acting rashly or impetuously.

"All haste is of the devil" was a medieval alchemical saying that was oft repeated by Jung. Something of this insight is reflected in our inborn suspicion of the person who rushes to judgment or who frequently waffles or changes her mind. As legal scholar Stephen Carter writes in *Integrity,* we suspect that she has not "engaged in the hard and deep discernment that makes integrity possible." Taking the time to reflect, says Carter, is crucial to integrity. Not to do so, he believes, enhances the possibility of doing wrong as "the fact remains that we cannot know what is right unless we think about it first."

The virtue inherent in solitude is evident in the great respect that, throughout history, has been accorded monks, hermits, nuns, and other recluses. As author and "semi-hermit" Peter France points out in the book *Hermits,* one of the ironies of the human situation is that "those who have chosen to live outside society have always been eagerly sought out for advice on how to live within it." Hermits have earned their reputation for wisdom, he observes, not merely for their heroic asceticism but for "insight into the ways of the world." Cultivating the habit of solitude, like exercising or eating a healthy diet, however, is available to us all. A part of each of us is a natural hermit, says Pir Vilayat in *Awakening,* a mystical being dedicated to the pursuit of spiritual truth and awakening. For while friendship and familial relationships are essential to psychological wholeness, the intimacies of our lives must be balanced by the capacity to be alone in order to grow close to the soulful being harbored within ourselves. To spend hours of time immersed in reading, listening to music, or contemplation strengthens this connection to the silent side of our personalities, and can be counted a valuable source of inner power and inspiration.

I still carry within me a memory from childhood, for instance, of my mother as she headed off for her solitary walks in the country—a stone in her pocket that she turned over and over along with her thoughts. A healer I know speaks of the joy that comes to her in the silence of an evening spent simply watching her candles flicker in the dusk—a practice that replenishes the energy she draws upon in her work. Another woman I know takes periodic retreats to a health center set deep in the Virginia woods. There she hikes, broods, and disentangles the moral complexities of her life. Yet another businessman I know is an avid sailor. Out on the vast stretches of high sea, with not even the rim of a horizon to distract him, he thinks: about his intimate relationships, business transactions, and the decisions he faces as a parent. Though not particularly devout or religious, this habit adds a dimension of depth and consideration to his character that sets him apart from others.

There is a Sufi practice called *Muhasaba,* or examination of conscience, that stresses the significance of moral reflection to spiritual unfoldment. Traditionally, it is a practice undertaken at the outset of a retreat, as before the retreatant can embark on her inner pilgrimage she must engage in a period of intense inner scrutiny. The practice of *Muhasaba* symbolizes the link between our day-to-day lives and the spiritual ideals toward which we aspire. It involves an honest reevaluation of one's priorities in life, stepping back and surveying the motives and intentions underlying activities and personal involvements. The soul-searching of *Muhasaba* reveals where our energy has been directed, and whether or not we have been living up to those ideals we hold most precious. It is a time to practice "deathbed wisdom"—looking back over our lives as if we were dying, reviewing the decisions we have made and whether the choices we have arrived at truly reflect how we would have *really* wanted to live our lives. Our values, as well, come under scrutiny, as we list those things that really matter to us and determine whether we have been living according to the principles we hold most dear—or to those that satisfy conventional expectations.

Sorting through the true and the false values of our lives is a task well suited to Mercury as it backtracks through the zodiac, giving us the time we need to ponder decisions and dilemmas. Surrounding ourselves with a zone of silence, we move away from ordinary surface habits of thought toward the very source of thought itself, the seat of intuition. Here we encounter what astrologer Dana Gerhardt has called "the listening twin"—our twin soul, our inner self who holds the answers to our most urgent questions. In this intuitive, less logical frame of mind, we tap into a stream of inspirational reflection more closely akin to the philosophers or thinkers of old. It is in this state that Jesus spent forty days in the wilderness, wrestling with the devil and his temptations, or Buddha while he was meditating beneath a tree on the banks of the Nairanjana River contemplating the human condition and seeking enlightenment.

Modern life may not accord us the freedom to go into the wilderness, as did Jesus or Buddha. But if we can find a way to create the right conditions for it, the first phase of Mercury retrograde lends itself to accessing a nonverbal level of consciousness that hums just beneath the threshold of everyday life. The meditation teacher Pir Vilayat Inayat Khan, for instance, often instructs his students in a meditation exercise to close off the outer senses and awaken the inner senses. In this exercise fingers are placed over the eyes, ears, and mouth in order to better listen to what he has called the "thinking of the Universe." As we breathe in, our hands are loosely placed over these openings; as we hold our breath, we press down tightly—and listen to the deep roar of silence that, like an invisible ocean of sound waves and light, washes over us. After lifting our fingers and breathing out, we repeat the practice again.

Different from the "acquired" knowledge culled from our experience of everyday life, this exercise connects us to a form of "revealed" knowledge—the ever-creative thoughts that arise untainted from the well of the cosmos. This fountain flow of visionary thoughts and spiritual currents is not our own. It comes from somewhere else, and, briefly, we touch a mysterious dimension far greater than ordinary reality. As if from a great height or a great depth, we see things differently. For when we look at life as if we are seeing through the eyes of God, writes Pir Vilayat in *Awakening,* "we access an inborn, intuitive, revealed knowledge that exists irrespective of the human condition." Thus through silent contemplation or intense concentration, we can learn to shift outer consciousness through inner attunement. From the broadened, expanded vantage point of the "thinking of the Universe," we can focus on that which is *teleological,* or purposeful, within our life problems—shape shifting and transmuting dark energy into gold.

The old temptations of the world—among them greed, envy, and power—still exist, as they did for the great religious prophets. But to immerse the mind in the deep waters of meditative reflection is to allow the "still small voice" within to speak—and to voice its guidance on the conflicts we are struggling to resolve. For though it may seem small in comparison to the loud clamoring of the outside world, it arises from a place of great wisdom and power, blessing every decision we make and gracing every word we speak with the perfume of silence.

8.

THE MUSE OF MEMORY

You will wake, and remember, and understand.
—Robert Browning, *Evelyn Hope*

Life can only be understood backwards, but it
must be lived forwards.
—Søren Kierkegaard, *Life*

As we enter into a more reflective frame of mind under the influence of Mercury retrograde, memories of the past may begin to float to the surface. For the greater part of our lives, we are compelled by the pull of the future. Yet at other moments the past impedes this forward flow, tugging at us like a whirlpool or a crosscurrent that suddenly swamps our souls. Especially during Mercury retrograde, we may feel in the grip of memory, pulled by the hands of history downward into the well of remembrance. Like the "awesome Charybdis," the treacherous whirlpool in *The Odyssey* that "gulps the dark water down," there are times in life when that which has gone before overwhelms us with emotion.

As astrologers have observed, the cycle of Mercury retrograde is so named because the planet slows down and retraces its steps throughout the portion of the zodiac it has just traversed. Symbolically, then, we are being asked to revisit our past, to go back and reconsider and review recent actions. Sometimes this is necessary because, in our haste to achieve something or get somewhere, we may have overlooked an important step or forgotten something along the way. Or, it may be that a part of our past has gone too long untended, and calls now for our consideration: the unresolved romantic relationship, the unfinished book, the friend with whom we have fallen out of touch. Even those things that lie in the far distant past may suddenly resurface, whether a long-buried childhood memory or, beyond that, a distant echo from another lifetime. Whatever unresolved business haunts us is likely to intrude into the present during Mercury retrograde.

The intermittent backward tracking of Mercury retrograde, in fact, is like life itself. Rarely do we proceed in a straight line from milestone to milestone. Rather, our paths are labyrinthine as, moving forward, we are waylaid by the past and must circle back, down, and around again. Like Odysseus, we yearn for a smooth journey only to find that the gods continually impede our way with strange detours. In *Retrograde Planets,* astrologer Erin Sullivan includes computerized images of the cycles of Mercury retrograde over a period of time; the pattern it traces resembles circles of intricate and interlocking loops. It is a stellar design that accurately pictures the circuitous spiraling that marks the human condition. Today, many modern-day pilgrims experience this when they undertake the spiritual practice of walking the labyrinth. This ancient symbol of a single circular path that leads to the center had been found in cultures around the world, from the thirteenth-century Cathedral of Chartres to the legendary Cretan labyrinth of Knossos. As they follow reproductions of this winding pathway laid out in churches or grassy fields, participants find themselves circling around and around—veering in close to the middle, then just as suddenly on the outer rim. As they finally reach the heart of this twisting, in-and-out pattern, seekers discover that the labyrinth is a metaphor for the journey of life. Less like a straight path to a fixed goal, the labyrinth is a physical experience of life as a meandering odyssey that weaves across many dimensions of time—past, present, and future—leading ultimately to a sacred inner center.

To find the psychological meaning in the labyrinthine, forward and backward motions of the planet Mercury, Sullivan turns in her book to the Greek myth of Prometheus and Epimetheus. When Mercury is moving direct, as it is about 80 percent of the time, writes Sullivan, "the mind operates on a very functional level," and energy is expended in productive action. Sullivan names this the *Promethean* phase of the cycle, after the Greek hero who stole fire from the gods to give to humankind, directly disobeying Zeus's orders. Prometheus, which means "forethought," represents the part of us that acts on gut instinct, without regard to consequences. In this frame of mind we are more likely to jump quickly into action, implementing our plans without thinking things through.

When Mercury is retrograde, however, this principle is reversed and we are more likely to stop and think before we act. Sullivan calls this the *Epimethean* phase, named after Prometheus's brother Epimetheus. As the story goes, Zeus punished humankind for stealing fire from the gods by creating Pandora, whom he gave as a wife to Epimetheus along with a chest. Although forewarned by Epimetheus not to touch the chest, Pandora, consumed with curiosity, opened the lid anyway. In doing so she released the terrible spites of old age, illness, death, vice, and passion that had been sent by Zeus as afflictions. All that remained was hope, the stubborn faith that, despite the chaos in the world, all will be well. Thus Epimetheus, or "afterthought," occurs during that phase of the retrograde cycle when the mind, recoiling from the aftereffects of hasty action, turns inward, digesting the meaning of all that has gone before.

In the myth of the brothers Epimetheus and Prometheus, as in the cycles of Mercury, we see the mental two-step familiar to all of us: the forward and backward motion that takes place as we leap eagerly into life with little hesitation, then are forced by circumstance to reflect upon the consequences of our actions. The heroic war of the *Iliad,* for instance, is followed by the dreamier, more retrospective tale of Odysseus on his homeward-bound journey. *The Odyssey* is embroidered with Odysseus's musings as he retells his past adventures, brooding over his life. Any kind of biography or retrospective, such as Jung's *Memories, Dreams, Reflections* or *The Odyssey,* is a literary form of Mercury retrograde expression.

Musings, reveries, recollections: During Mercury retrograde the muse of memory is awakened. For all those who are philosophically minded, or who seek to know life's truths, there comes a time when the need to pause and reflect upon the road traveled thus far grows strong. Nostalgia for the past, for home and history and heritage, sweeps our souls with longing. Here Mercury-Hermes emerges in his role as psychopomp, leading us down into the underworld of the unconscious. Delving into the basements of our psyches, we examine the relics and remnants of our previous lives, seeking clues to the mystery of our identity—a keepsake that reveals the plotline we are living or the secret to our true identity.

This rhythmic return to the past happens to all of us periodically over the course of a lifetime. After leaving home as young people, we may suddenly feel the need to return to our family to hear once again the stories of ourselves as children, or to touch the mementoes left behind. After the end of a marriage or relationship, we may unexpectedly find ourselves immersed in thoughts of our former partner, as we contemplate the history of what worked and what went wrong. When our child grows up and leaves home for the adult world, we may pine for the toddler we once cradled on our lap. Upon quitting a job, we may look back and ponder the achievements and losses, the relationships made and connections broken. Entering old age, we may find ourselves revisiting the scenes of our youth.

Remembering the past, digesting our previous actions, and turning over and over in our minds old and unresolved issues is the core of soul work. In the Greek myth of Mnemosyne, the muse of memory was a titan—a giant goddess who mothered nine muses of her own, each one a talent. This tells us that memory is a formidable force to be reckoned with, a creative, shaping power not to be ignored. For without the important work of memory we would have no history to call our own, no singular narrative to define us as unique individuals. It is the past that shows us where we have been, in order to measure for us how far we have come. The person with no material souvenir or connecting link to his past, for instance, suffers a terrible wound of loss. Why else would people mourn with such intensity when fire, flood, or political disaster robs them of the inconsequential material artifacts of their lives—the photo albums, worn stuffed animals, or cheap trinkets—that are symbolic of old loves, small children, and dead relatives?

Memory is important because without it we would live outside the boundaries of story. And in story, says Jungian psychologist James Hillman in *Healing Fiction*, is soul—and in soul is story. Memory is the mythmaker, writes Jungian Ginette Paris in *Pagan Grace: Dionysus, Hermes, and Goddess Memory in Daily Life*, that weaves the fabric of our lives. For in reciting and recollecting the many incidents, episodes, and challenges we have encountered along the way, we become the bards of our own ballads or epics—the ordinary stories that are the extraordinary stuff of myth. In the process, we reawaken an ancient part of our human heritage—the ability to mythologize. "Unfortunately, the mythic side of man is given short shrift nowadays," wrote Jung in *Memories, Dreams, Reflections*. "He can no longer create fables. As a result, a great deal escapes him; for it is important and salutary to speak also of incomprehensible things. Such talk is like the telling of a good ghost story, as we sit by the fireside and smoke a good pipe."

CREATING A SOUL MEMOIR

There is no better time to sift through the treasures of the past than the days and weeks of Mercury retrograde. Like Jung, we can metaphorically "sit by the fireside" and ponder the mysterious myth that is our life. As we put aside the cares of the day, it is a time to dance backward along memory lane. Nothing, in fact, adds depth to

our present-day lives like an evening or an afternoon spent reminiscing about who we once were and what we once did. The older we get,
in fact, the better the old times become.

Often, it seems, without even trying old friends from the past
tend to call or come to town while Mercury is retrograde. When an
old boyfriend once visited, for instance, we spent a lovely evening recalling our madcap romance in all its youthful beauty. Another time,
a close girlfriend came to town whom I had not seen for nearly fifteen
years, and we spent a halcyon few days going back over the years
when we had lived close by to each other, and catching up on the
lives of our children. My Sufi friends and I love nothing better than
to recall our early communal days and to tell amusing stories of Pir
Vilayat, our teacher. Now that my children are grown, it is a joy when
they return home for visits to gather with their old friends and remi-
nisce about the times when they were young. Like a sponge that expands when dipped into water, the original experience grows in the
interlude between past and present, enriched by the fullness of time.

But the past enters the present in other ways; once, as a girl, I remember my mother being suddenly gripped by memories of her de-
ceased father. Puzzled, she realized that the vivid memories were
surfacing because of a new brand of ice cream she'd been eating—
the taste had reminded her of ice cream she had enjoyed with her father as a young girl. As such incidents show, we don't have to visit a
psychic to realize that the past is never very distant. Rather, the invisible world touches the visible through the soft, angel brush of the
senses: like Proust's infamous madeleine that began his classic novel,
a whiff of perfume, a touch of fabric, an overheard refrain of music,
or the sight of a person walking in just such a certain way down the
sidewalk can resurrect history all in a flash. We can think of these uncanny instances as the touch of Hermes, as he beckons us back
through the preceding chapters of our life story.

Indeed, each of us, writes James Hillman in his extraordinary
book *Healing Fiction,* has our own personal memorial hall, rich with
the memorabilia of images, the monuments, icons, and temples of the

past. In my own memorial hall, for instance, I see the weathered foundation of an old building hidden deep in the woods by my house; the high school football field in the crisp fall air dancing with orange-clad cheerleaders; the overstuffed red chair that was my first real piece of furniture. These memories are much more than the isolated images they appear to be. Rather, they are like archaeological fragments that, when excavated, restore something precious that had been broken apart by the rushing flood of time. These early images, artifacts of memory, begin the narrative epic that forms the trajectory of our lives. From them, history is rebuilt and meaning is constructed.

This, of course, is the work of therapy: the retelling of our past, of what has gone before, with an eye to how the story of the past has shaped the story of the present. As anyone who has ever been in therapy can attest, the process begins with the chain of events we relate to our psychotherapist. Usually beginning with our parents, we trace a line directly leading to the dramatic event or turning point that brought us to therapy in the first place. Story, in fact, says Hillman, is central to the healing process of psychotherapy. Those born with a sense of story built in from childhood, he writes, are in better shape than those who have not had stories, nor had them read, or acted out, nor have made them up. For through the sensory, subjective act of reminiscence, maintains Hillman, the linear, objective facts of our lives—the who, what, when, and where—are transformed into soul stories. The kind of biography we construct for ourselves while in therapy has less to do with the outer facts of our lives and more with our inner, subjective experiences. The dreams and fantasies that arise as a response to events form their own internal narrative, giving definition and distinction to the personality. Like the pearl formed around the grit of sand in the oyster, our stories are often formed in the liquid ocean of feeling and emotion.

Indeed, how we craft the literary "fiction" of our lives to our therapists or to ourselves, maintains Hillman, reveals the facts of our soul, or our "soul histories." In other words, as most writers know, *how* the story is told is as important as *what* the story really is. "The way we imagine our lives is the way we are going to go on living our lives," writes Hillman in *Healing Fiction,* "for the manner in which we tell ourselves about what is going on is the genre through which events becomes experiences." Regarding history from the viewpoint of the soul, explains Hillman, reveals the god, myth, or archetype that shines through as "there is a God in our tellings and . . . this God shapes the words . . ."

RITUALS OF REMEMBRANCE

The god behind the act of storytelling, of translating the facts of our lives into fiction, is Hermes. Not only does Hermes guide us down into the underworld of our psyches, it is Hermes who recounts the story afterwards. It is Hermes' function, in other words, to interpret for us the essential meaning to be found in the inchoate dreams and longings of our inner lives—to craft the story out of the remnants, the bits and pieces of feelings and perceptions. Hence the term *hermeneutics,* to interpret, to extract the essential core of something. Without meaning, the experiences of our lives would simply roll by, carrying us along in the wake of time. In my own life, for instance, a significant dream will often accompany a life event. Something about that dream, whether an evocative image or a moody landscape, adds another dimension to what I have just experienced, whether a disagreement with one of my sons or an essay I am writing. The dream itself, in a sense, acts as narrator of my life's film. Journaling about the dream, or discussing it with a therapist or friend, I discover even more insights into and revelations of the *real* story, the mythic plot, of my life.

Following the deeper layer of mythic commentary that runs like an underground river throughout our lives is one of the best uses of Mercury retrograde. But we have to take the time to explore it. I would not gain the gift in my dream, for instance, without devoting an hour in the morning to writing in my journal, sifting and sorting through the vision that visited me in the night like a miner panning silt for a nugget of gold. The narrative of our soul history cannot be uncovered on the fly, but only in the slow hours of solitude and reflection. Soul is not found before us, but behind us; thus we risk passing it by altogether unless we slow down, regress backward, and turn and take the long glance into the shadow cast behind us.

Had we more time for experiencing life in this way, digesting and ruminating over the parade of people and events through our lives, we might find the meaning in life so many of us seek. Everyone knows the peculiar sensation that comes from dashing from one thing to the next, and the intoxication that comes with the accompanying rush of the thrill. Yet it is also true that when things happen too quickly, something is lost, as if what has happened hadn't really occurred at all. And indeed, nothing really penetrates the surface of our lives if we have not slowed down enough to experience it in retrospect, rewinding and replaying the event in our minds until its true import finally sinks in.

Mindless, automatic repetition of events in our mind's eye doesn't always bring awareness. But consciously recalling past circumstances, either alone with our conscience, or with a counselor or trusted friend, can help to integrate the meaning of a significant life event. As anthropologists Jan Clanton Collins and Thomas Gregor write in *Romantic Passion,* couples with enduring bonds enjoy telling and retelling the narrative of their relationship. "So deepening was the experience," write Collins and Gregor, "that . . . the narrative is . . . a way in which love partners enact their loving relationship." Whoever has been in love is familiar with the romantic ritual in which two people recount to each other the odd details of their first encounter: the place they met, the first exchange of a glance or the touch of hands, the flirtatious e-mails or postcards.

Those who work with victims of trauma know that repetition is an important part of the healing process; going back over a terrifying incident, again and again and again, helps to reweave the hole torn in the net of the psyche. Repetition is a sacred principle found in the ancient mysteries of the pre-Christian world. Rituals and ceremonies were based on the reassuring repetition of the solar cycles, as the Sun returned each year to the same place in the sky, or the annual round of the seasons. The Catholic mass is a re-presentation of the crucifixion and resurrection of Jesus Christ. For most artists, the secret of their craft lies in the art of going back over something—rewriting the sentence, repainting a scene, or repeating a refrain from a song—not until it is perfect, but until it assumes the shape or sound it is supposed to be. For it is through the process of regression and repetition that we find that niggling *something,* that essential missing ingredient, that through finding makes us feel right again.

When Mercury turns retrograde, we can be sure that something is calling to us from our past that we have forgotten to take care of along the way. Whether an unassimilated experience, an untended relationship, or a piece of ourselves, we are being asked to regress in the service of the soul—to journal and talk and meander and put it all down in story, legend, poem, and narrative. The retrograde and direct phases of Mercury retrograde offer us an ideal template for working with the backward and forward flow of time and experience through our lives. While Mercury is direct, we can feel free to "go with the flow," moving through life unrestricted by too much reflection or analysis. Yet when Mercury turns retrograde, backing up through the zodiac, we, too, can go backward.

We can begin the process of going back over our lives with an eye to story by examining the events of the past three months, the interval since the last retrograde phase. What have we overlooked? What calls to us for our attention? It may be a sibling relationship that we have neglected, yet that contains within it an important piece of our own family history. It might be that an intimate relationship with a loved one needs tending in the form of long, intimate conversations that give voice to the feelings that have silently accumulated during the past months. Looking back over the recent months, we may catch a glimpse of a lesson learned or wisdom gained. Or, we may wish to visit a family member we haven't seen for a long while—the grand-mother with memories to transmit, or the uncle we've never really come to know. Something in our distant ancestry may beckon, com-pelling us to explore our genealogy and track down the lost legends of our families. One of my richest experiences during a Mercury retro-grade phase took place during a long overdue family reunion during a Christmas holiday. As my sister and brothers and their respective families gathered together with my mother, we spent hours together simply talking, laughing, eating, and reminiscing about the past.

The call could come from within, as well, in the form of neglected feelings and moods that now beg for our interpretation and integra-tion into the daylight of consciousness. Mercury retrograde favors psychological and therapeutic work, either in the analyst's office milling and processing the emotional flotsam of our psyches, or in the privacy of our own journals as we write down our thoughts and musings. It is a favorable time to devote to dream work, perhaps get-ting together with a close friend to share our nocturnal wanderings. Often, friends of mine take time to clean out drawers and closets during Mercury retrograde. Sorting through boxes of old photos, go-ing through clothes and books and bills while musing over our past can be a practical, soothing counterpart to the inner work that is go-ing on inside at the same time.

When recollecting the past that has gone before, it is important to include the larger story of the collective history that encircles our own. Too often, the stories we tell about our lives occur in a vacuum empty of the politics, scandals, wars, scientific breakthroughs, and artistic achievements of the surrounding world culture. But these his-toric themes are the meta-themes of our individual lives. World War II imprinted my parents' lives, just as the antiwar demonstrations and civil-rights marches of the sixties shaped the political views I hold. Likewise the Internet and e-mail, the bungled presidential elections of 2000, or the terrible tragedy of the *Challenger* exploding in space are interwoven threads in the tapestry of my own evolving story. Though we may not even know them, the tragedies, loves, and victories of strangers on faraway continents somehow make their way into our own soul story. In realizing this, the short-lived chapters of our lives become part of a longer, more enduring book of history.

So compelling a force is the backward-flowing river of time that it can even take us out of the realm of our own life and into past lifetimes. "Not only do retrogrades cause the individual to regress back to yesterday, last month and last year," writes esoteric astrologer Martin Schulman in *Karmic Astrology,* "but they also induce regression back to former lives whose memories carry strongly into the present incarnation. These memories represent specific events or individuals that were meaningful enough in another life to still have a hold on the person now."

We do not have to believe in reincarnation to realize that time is multidimensional and that the old world lives and breathes in the world we live in now. Thus the spirits of the Native Americans sigh along the river where I walk daily, and ancient Babylonians are with me as I gaze up at the same stars they once studied. The presence of the deep, long-gone past is always in the present moment. Thoughts of karma, of the forces of fate and destiny that are carried over from lifetime to lifetime, imbue our remembrances with spiritual mystery. Through the lens of a long-ago culture—ancient Greece, China, or India—we are inititiated into a different way of seeing the story we live now. Even the roles we are cast in today take on imaginative depth when measured against a possible past lifetime: If I am a mother now, what kind of mother was I then? If an astrologer today— what kind of astrologer then? Reflecting upon our past in the pool of time, an older, ancestral self may emerge out of the depths—the forerunner of our modern-day persona.

Old people often make the mistake of becoming imprisoned by their memories, wrote Jung in *Memories, Dreams, Reflections,* becoming lost in their reconstruction of past events. But if introspection is reflective and translated into images, he says, then we see the purposeful "line" that leads from our lives into the world, and out of the world again. What Jung seems to be saying is that, from a soul perspective, our memories have a transformational purpose that gives our lives meaning. To be sure, we all wish to escape the painful confinement of the past, to hurtle forward into unfettered freedom. But the wisdom of Mercury retrograde, to paraphrase the Bible, says that to everything there is a time—a time to pause and look backward, and a time to move forward.

For turning backward in time—not in bitterness, but in thoughtful contemplation—allows us to pick up the thread of story that we may have dropped in our haste to get ahead and keep pace with everyone else. It allows us literally to "re-member," or bring together, all the pieces of our self that have been lost or scattered along the way. Through the process of repetition and remembrance, we become renewed. Reconnected to our past—whether long ago in a distant era, or twenty years ago in America—we locate ourselves along a continuum of time. Life becomes more interesting, and less dull. More important, we remember who we really are, and find the mysterious something that is living inside us—a precious piece of the larger story of humankind. That gift itself is the pure, unalloyed gold of myth and story: our own *Odyssey,* the journey of a lifetime preserved for us by Mercury the storyteller and the muse of memory.

9.

CREATIVE MAGIC

> ... the small boy is still around and possesses a creative
> life which I lack. But how can I make my way to it?
>
> —C. G. Jung, *Memories, Dreams, Reflections*

> It is Mercury who assists the artist in his discovery and
> travels into the unconscious to bring the results of
> subliminal conclusions to the conscious mind for
> articulation.
>
> —Erin Sullivan, *Retrograde Planets*

There is one true thing about Mercury retrograde—it is one of the most imaginative periods of the astrological year. The creative magic possible under this cycle, however, only works if we see the gold shining through the dross of annoying delays, petty breakdowns, and disturbances. If we can take Mercury's not-so-subtle hint and use the retrograde cycle to slow down, be still, reflect, and remember, if we can dance with the detours and take ourselves more lightly—then we can take flight on the wings of fancy. Everything about Mercury retrograde, from the altered sense of time to the boon of additional time, conspires to inspire the artist within.

Especially as Mercury begins to round the corner, entering the concluding ten-day phase of the retrograde cycle that begins with the "New Mercury"—the joining together of Mercury and the Sun at the same degree in the zodiac—a fresh wind picks up. We sail on the exhilarating currents of newfound ideas and inspirations. Perhaps our imagination has been piqued by a memory from the past, jolted by an unexpected event, animated by research into a period of history, or uplifted by our sojourn through the valley of inner solitude. Because the planet continues to backtrack, however, and we may still struggle with feelings of frustration. Two currents conflict: An influx of new energy clashes with a mood of "not yet." Like a scientist closeted in her lab on the verge of discovery, or a writer inspired by a book's unfoldment, it is a phase of near-complete ripeness. It is a time, in other words, to allow ourselves to freely explore our creative impulses, without allowing our conscious minds to interfere with anxious predictions of what the end result of those impulses might be.

As a writer, I have often found Mercury retrograde an ideal time to work. Due to the slowing down of life that frequently occurs—when meetings are canceled or projects are postponed—I have found that sometimes there is simply more time in the day to devote to my craft. Because the act of writing requires a more deliberate, measured pace in order to access a deeper layer of thinking, it harmonizes with the oftentimes leisurely rhythms of Mercury retrograde. Revisiting a creative project begun earlier then put away, however, is also favored during this three-week period. It is a good time for the refined work of attending to detail and polishing a project to perfection—revising, rewriting, or editing. Astrologer and scholar Robert Schmidt, who translates ancient astrological texts, finds that the triple three-week retrograde periods throughout the year have always proved favorable to his backward-looking historical research into past civilizations.

But it is the kind of creativity inspired by the play of life that proves fertile ground right now—those projects that we undertake simply because they please our souls, rather than for any specific outcome. Reverie, daydreaming, musing, and playing around with ideas—these are the artist's tasks during Mercury retrograde. Rather than a finished product, we can turn our attention toward cultivating an atmosphere of invitation, experimentation, and receptivity—the fertile soil in which creativity can take root and blossom.

As every creator is aware, for instance, there exist many stages in the inventive process. Honing a book, painting, software program, signature dish, or piece of music into its final form is the last in a series of steps. It is the sizzling flash of insight, however—the "aha!" moment—that first sparks the creative process. Like Athena, the goddess who sprang fully blown from the head of Zeus, inspirations sometimes pop into our heads with godlike power. Paradoxically, such godlike moments tend to occur when we are in a state of light bemusement—as a child at play with clay, colors, and paper—rather than when we are seriously intent on our work.

The famous story of Archimedes of Syracuse, as Erin Sullivan points out, illustrates the link between creativity and a relaxed state of mind. As advisor to King Hiero II, Archimedes was trying to determine whether the king's crown was pure gold or alloyed with silver. Exhausted from his logical and analytical efforts, he drew himself a bath. As he was submerged in the warm water, the truth suddenly emerged, and he shouted the now famous "heureka!" having "found" the theory of displacement. A more modern example of this principle can be found in the delightful movie *The Muse.* In it, the actress Sharon Stone plays the role of a goddesslike *inspiratrice,* or inspirer, who helps a blocked Hollywood writer rediscover his genius by taking him on childlike outings to places like the aquarium.

The whimsical, colorful, and capricious muse played by Stone is an accurate personification of Mercury retrograde's tricksterish talent for the creative breakthrough. Indeed, astrologers say that the planet Mercury is in its feminine incarnation when retrograde. A copperplate engraving from 1563 pictures the rare figure of a female Mercury, bare breasted and holding a caduceus, or wand, in her right hand. She is surrounded by what are called in astrological lore "the children of the planet": a sculptor, painter, cosmographer, teacher, and builder as well as images of the seven liberal arts—arithmetic, astrology, dialectics, geometry, grammar, music, and rhetoric. Likewise a sixteenth-century woodcut *Mercury and the Artisans* shows the "children" of the planet bent over their worktables, wholly absorbed in writing, drawing, and making music. These medieval images illustrate the quest for knowledge that the planet signifies.

The desire to know and be creative is fueled by curiosity; of all the gods and goddesses, Hermes was perhaps the most inquisitive. His unabashed precociousness was evident from the first moment of his birth. Stumbling across the ordinary tortoise, Hermes asked himself, "What if?" quickly reimagining the lowly turtle's shell as a sophisticated lyre. All the elements of this story speak to Hermes' fanciful, imaginal capacities. From a foolish, childish prank emerged something never before seen—a musical instrument whose notes have enchanted humankind for centuries. It was in exchange for the lyre that Apollo gifted Hermes with divinatory powers, and indeed, magic and the creative process are often seen as synonymous acts of mysterious power. The notion that something suddenly appears out of nothing, as if out of thin air, has always fascinated and beguiled. "Where did that idea come from?" would-be artists wonder of their favorite author, musician, or painter, just as Apollo wondered in awe at the inventive prowess of his little brother.

As Mercury is the planet that rules curiosity and knowledge, we may find ourselves during the planet's retrograde cycle gripped by an insatiable desire to delve into the study of a new subject or artistic craft—a sure sign that our creative spirit has claimed our attention. The excited student we once were may resurface, eager to tackle a new lesson in math or history—or to compose a poem or improvise a new recipe. If we follow Mercury-Hermes' trail through the zodiac, we may, in fact, arrive at the source-spring of inspiration. Weaving backward through the mythic time zone of Mercury retrograde, we may rediscover our own latent curiosity and desire to know more about the world we live in, tapping into sparkling underground reservoirs of creativity.

THE SOUL AT PLAY

Children at play rarely consider what they are doing as finished projects "for sale" or display. Rather, they are wholly absorbed in the dizzying pleasure of the act itself, delighting in the feel of paper and the smell of paint, or squealing in happiness as they squish clay into whimsical shapes that please only themselves. For the longest time, my creativity was sparked by watching my own children after school as they raced home to embrace the freedom of their unscheduled hours. Watching them carry on imaginary conversations with plastic figures and stuffed animals, bike endlessly up and down the street, build intricate Lego castles, structure elaborate forts out of pillows and chairs and practice skateboard flips, I basked in their joyful abandon to the moment. If we can remember back to the child we once were, to our precocious and uninhibited self, that kind of ever-regenerative creative spirit can be rekindled.

Something about the child within us, in fact, holds the key to our artistic selves. Often as we grow older, the weight of life buries that once happily playful part of us beneath a burden of responsibility and care. We grow distant from spontaneity, laughter, and a sense of life as an adventure of discovery. Fixated on goals and bottom lines, dependent on other people's opinions, our creative energy is bled away until we become worried shadows of our former selves. Often when I find my own well of inspiration depleted, or feel discouraged, I take a break from work and try to re-create my mood as a child. As I recall the quixotic, secret stories I used to make up in my mind, my dreamy fascination with books and history, and my living connection to fairies in flowers and spirits in attics, the waters begin to well up from below once again.

The great psychologist Carl Jung, in fact, healed himself of the trauma he endured when he made his historic break with Sigmund Freud by reliving certain key scenes in his childhood. Disoriented and "suspended in mid-air" without Freud's support, as he writes in *Memories, Dreams, Reflections,* Jung found himself at the threshold of a new life. Though he felt compelled to find a new way of working with his patients without relying on previous theories, he still did not know yet what that way would look like. Giving himself up to his dreams, Jung turned to his unconscious for guidance. Deeply disturbed and under a constant inner pressure, Jung made the radical decision that "since I know nothing at all, I shall simply do whatever occurs to me."

To Jung's great surprise, the first thing to surface was a childhood memory from his tenth or eleventh year when he had enjoyed playing with building blocks. As he recalled constructing little houses and castles with gates and vaults using mud and stones, he was flooded with emotion. "Aha," Jung said to himself, "there is still life in these things. The small boy is still around, and possesses a creative life which I lack. But how can I make my way to it?" Amazingly, Jung did not just simply reimagine himself as his former eleven-year-old self. He actually acted it out in real life, going outside for a period of time every day and making buildings and villages out of mud and bricks. Here is how he describes this process in his memoir:

> . . . as a grown man it seemed impossible to me that I should be able to bridge the distance from the present back to my eleventh year. Yet if I wanted to re-establish contact with that period, I had no choice but to return to it and take up once more that child's life with his childish games. This moment was a turning point in my fate, but I gave into it only after endless resistances and with a sense of resignation. For it was a painfully humiliating experience to realize that there was nothing to be done except play childish games.
>
> Nevertheless, I began accumulating suitable stones, gathering them partly from the lake shore and partly from the water. And I started building: cottages, a castle, a whole village. The church was still missing, so I made a square building with a hexagonal drum on top of it, and a dome . . . I went on with my building game after the noon meal every day, whenever the weather permitted. As soon as I was through eating, I began playing, and continued to do so until the patients arrived; and if I was finished with my work early enough in the evening, I went back to building. In the course of this activity my thoughts clarified . . . Naturally, I thought about the signif-

icance of what I was doing, and asked myself, "Now, really, what are you about? You are building a small town, and doing it as if it were a rite!" I had no answer to my question, only the inner certainty that I was on the way to discovering my own myth. For the building game was only a beginning. It released a stream of fantasies which I later carefully wrote down.

Jung's experience proved pivotal and throughout the rest of his life whenever he came to a brick wall he painted a picture or carved in stone. One of the twentieth-century's most original thinkers, Jung's purely spontaneous acts of creativity functioned as rites of entry for his most important books and ideas. In it we see the origin of the Jungian technique of active imagination—spontaneously drawing or painting our unconscious fantasies. In reading this account from Jung's life, I was struck by the faith he placed in his intuitive inner guidance, to the point that he was able to dismantle his adult defenses and play like a child. His capacity to do so proved life-changing, as it was Jung's childlike trust in the unconscious that served to keep him on his life course.

So often we are exhorted to "live our dream" or find our life purpose, yet rarely are we given good advice as to how to go about it. Even the simple statement to "do what you love best" overlooks the fact that discovering what that is may require long-drawn-out periods of not-knowing. To find our dream sometimes necessitates a casual, offhanded playing around with ideas, like tossing a ball back and forth or digging in the dirt. During this time we may go through an open-ended, trial-and-error period as we literally fool and mess around a lot, mixing things up. We may travel a lot of roads, only to find that some of them may lead to dead ends. Like the ancient alchemists who stirred and cooked elixirs to distill gold out of lead, or a child mingling finger paints, the creative process is sometimes a messy, unpredictable affair. Still, as Jung's story indicates, each new beginning—whether creative or psychological—arises from being open-minded enough to allow some unexpected element into our lives, without knowing exactly where it may lead us.

Play, some say, created the very world we live in. Some mystical traditions hold that God created the world because He—or She—was bored being alone, and needed a diversion. In her book *Creation Myths,* Jungian psychologist Marie-Louise Von Franz relates many creation stories that are remarkable in their whimsy. In one Eskimo myth, the creator-god Father Raven plants herbs and flowers on the new earth. As he opened a pod one day, to his great bewilderment a human being popped out. As Von Franz recounts the myth, "Then Father Raven laughed heartily and said: 'Well, well, you are an odd creature! I never saw anything like you!' Then he laughed again and added: 'I, myself, planted this pod, but I did not know what would come out of it.' "

In her discussion of art and the imagination, Von Franz talks about the importance of accidental actions in the creative process—the smudge in the painting, or the dent in the pot that accidentally reveals the true design. Many inventions throughout history, writes Von Franz, have been the result of an accident, "through someone playing with an object and then suddenly getting an idea." Like a lawyer on *Oprah* who turned his unusual passion for baking cakes into a profitable bakery selling his luscious creations, what we stumble across as if by accident may be where our real genius lies. Emerging theories in science show that accidental creativity plays the same role in nature. According to the theory of self-organized criticality, nature evolves through unpredictable events and catastrophes. Life as we see it today, writes physicist Per Bak in *How Nature Works* "is just one very unlikely outcome among a myriad other equally unlikely possibilities."

Potentiality, possibility, probability—this is the stuff of the universe, and the stuff of creativity as well. "I dwell in Possibility—" wrote the poet Emily Dickenson. "A fairer House than Prose—More numerous of Windows—Superior for Doors—" The effortless ease of jazz musicians as they play off each other's rhythms, composing a piece of music that floats on the beauty of the passing moment, or a free-ranging philosophical dialogue are examples of innovative thought in action. Like fanciful embroidery, it is a state of mind that is less linear and more dreamlike than the goal-oriented, logical, and rational approach we are called upon to take in the everyday world.

Aside from creating the conditions that lead to both great and small works of art, however, playing is also the secret to a well-lived, happy life. As adults, we rarely experience the juicy surge of physical energy that comes from simply running wild through the open grass, or flinging ourselves with abandon into a pool of water on a hot summer's day. Games, jokes, even good-natured pranks, evoke Hermes' creative trickster spirit. Mercury retrograde is one of the best times for letting down our guard and learning about life through the pleasures of the senses. Rolling around on the floor with one's dogs, playing jacks with a child, enjoying a good game of cards or dancing the night away all serve to loosen our outmoded ways of thinking, jogging our inhibitions and putting us in a creative frame of mind.

Even slacking off and doing nothing have their creative virtues while Mercury is slowly retreating through the zodiac. Though they may not speak of it very often, most artists know the value that comes from just sitting in a chair and staring out the window, listening to music, taking a nap, going to three movies in a row, or even drinking wine and watching television in the afternoon. One of the classics of modern literature, *Zorba the Greek,* by Nikos Katzanzakis, tells the story of the young, intellectual writer whose somewhat detached and objective perspective on life is overturned through his encounter with the irresponsible, irrepressible Greek workman, Zorba.

Zorba is the prototype of the playful artist wholly abandoned to the unpredictable and wildish forces of life. With his stringed *santuri* and unabashed desires for dancing, feasting, lovemaking, and drinking, Zorba charms and enchants his bookish companion. In this passage, we get a glimpse of Zorba's tricksterish, alive spirit: "As he spoke, his imagination kindled. His eyes sparkled, and like a poet in the burning second of creation, Zorba soared to heights where fiction and truth mingle and resemble each other, like sisters."

No livelier words were written of the inspiration that can be sparked when we free the artist within to play with abandon in the fields of imagination. For though it may seem frivolous to some, the power of play in the creative process functions as a screen over an important process taking place in the unconscious mind. Outwardly preoccupied, we are inwardly hard at work, as thoughts and ideas take root, germinating in the milieu of our imaginations. Out of sight from the prying eyes of the world—or our own harsh inner critic—a magical alchemy takes place as our minds are transformed below ground in the soul's creative depths.

THE PSYCHIC IMAGINATION

One of Hermes-Mercury's most significant mythic roles was as the wise Magus, or magician. Particularly during Mercury's retrograde phase, the veil between the seen and the unseen world thins, and the spirit of guidance shines out more clearly. Thus oracular techniques that draw inspiration from another dimension—whether throwing the I Ching, concentrating on a tarot-card spread, consulting our dreams, or even visiting a well-regarded psychic or astrologer—are strengthened at this time. The guidance we receive from these techniques may help us to resolve problems in our lives—but only if we are not too literal. Often, those who consult psychics or astrologers make the mistake of taking their advice too concretely. Rather, if we take a more artistic approach, we can use these consultations to jog our minds in a way that expands and reconfigures our habitual viewpoints. In this way, techniques of magic and prophecy help to get us out of the rut of our fixed ways of thinking, stimulating our creative imagination by sparking new ideas and engendering fresh perspectives.

Because Mercury retrograde brings out the more intuitive, imaginative side of our thinking, we may be more receptive to the psychic dimensions than during other time periods. In essence, magical techniques are a kind of medium that speak in the language of images, symbols, and signs. Like Mercury, images mediate between this world and the beyond. In the same way poetry or a painting can give expression to a wordless feeling, images communicate messages that can bring clarity and meaning where there was doubt and confusion. The dots that don't connect suddenly assume a beautiful and coherent pattern. Just as the stars in the heavens have guided sailors across the midnight sea, so the images we receive from a psychic source are signs that reveal the tracks of our destinies.

Often, techniques such as astrology or channeling are referred to as psychic "arts." And in fact, the language of imagery used by those in the occult professions is the same medium used by artists and musicians and writers to tap into the vein of creativity. That is why, when we visit a good psychic or insightful astrologer, we often come away feeling revivified. It is because the rich, visual images that are used to convey information provoke within us a creative change of mind—in exactly the same way a master painting, brilliant movie, edgy rap song, or poignant novel also changes our minds and gives us a new way of seeing things. And when we look at life differently, we envision our future differently. Suddenly, we are aware of possibilities and potentials that we had never seen before. Something mysterious transpires, our mental outlook shifts, and, as if by magic, so does our future. Whether contacting a departed spirit or being inspired by a poem, what really matters is that we have opened the door into the enchanted world of our own psychic imagination. For though we may not think of it in just this way, our imaginations are a living, teeming realm of spirits, people, animals, trees, monsters, stars, oceans, angels, prophets, wizards, and saints.

We cannot retrieve the inspirational messages lying in wait for us in our psychic imagination, however, unless we make a point of asking for them. Just as we must formulate a question in our minds before we throw the I Ching or ask for a dream from our unconscious before we fall asleep, we have to shift out of our ordinary frame of mind into a different mode of thinking. Usually, this requires a descent into the deeper layers of our psyche. The story of Odysseus in the Kingdom of the Dead illustrates this principle.

Midway through *The Odyssey,* Odysseus is instructed by the beautiful nymph Circe—at whose palace he has pleasantly tarried for a year—that before he can return home he must pay a visit to the blind seer Tiresias in the watery realm beneath the sea. "You think we are headed home, our own dear land?" announces Odysseus to his startled men, as Robert Fagles translates this moment. "Well, Circe sets us a rather different course . . . down to the House of Death and the awesome one, Persephone, there to consult the ghost of Tiresias, seer of Thebes."

Blown by the North Wind, Odysseus finally reaches that place "where the roads of the world grew dark," and "the eye of the Sun can never flash his rays." Gathering his courage, Odysseus digs a trench, then pours out libations of milk and honey, wine and water, as he has been told to do by Circe. Invoked by Odysseus's sacrifices, the "ghosts of the dead and gone" flock toward him. Tiresias, too, finally, makes his appearance, prophesying that Odysseus will complete his journey and reach the shores of Ithaca, where he will eventually die a peaceful death in old age—but only after enduring extreme tragedy and further hardship. Next, wave upon wave of shades—fallen heroes from the battlefields of Troy and the gods and goddesses of an older age—step forth and tell their stories. His dead mother gives him news of his still-living wife, Penelope, and his son and father, comforting him with a glimpse of the home he yearns to reach. Through it all, Odysseus's emotions are touched, as he is moved to tears of sorrow, feelings of anger, and waves of grief as he encounters the full spectrum of human experience in the Kingdom of the Dead.

In *The Odyssey* this entire passage—Tiresias's prophecies and the gods' exploits—is retold by Odysseus in flashback in the court of Queen Arete, where he has landed just before he reaches Ithaca. Enthralled, the assembled court of the queen is held rapt by his tales— a story within a story within a story. So talented is Odysseus at spinning and weaving his out-of-this-world experiences that he is referred to in *The Odyssey* as "the old master of stories." I relate this episode from Homer's classic epic because it shows how creativity is a kind of channeling from a remote and silent landscape within ourselves. Just as Odysseus was able to hear the stories of the shades of the underworld by invoking them to come forward and speak, only to retell them later as stories, so, too, can we use certain psychic techniques to invite images from the collective unconscious to speak to us. So numinous with enchantment is this realm that we have only to retrieve a shard or fragment from it to receive enough inspiration to write a story, paint a picture, or compose a song.

History is laced with stories of the link between creativity, the occult, and the imaginal unconscious. The great jazz musician John Coltrane was said to have been an avid student of mysticism and astrology, inspired by the planets in his creative pursuits. In *Her Husband: Hughes and Plath—A Marriage,* biographer Diane Middlebrook reveals that the English poet Ted Hughes had an intense fascination for the occult, particularly astrology. According to Hughes's friend Lucas Myers, writes Middlebrook, the poet "loved the opulent lexicon of symbols, the convergences, oppositions, planetary solar and lunar influences, the cusps and houses with which it organized a description of the human character and destiny." Yet, relates Myers, Hughes was not a literalist about it. "Ted saw astrology not as a science but as an instrument for the vivid expression of intuitive insights." So important was Hughes's natal birth chart to him in shaping his life myth that Middlebrook includes a reproduction and detailed discussion of it in her book.

Dreams, too, nurtured the direction of Hughes's literary career. As a student in English literature at Cambridge, writes Middlebrook, he once dreamed that a door opened and a man entered in the shape of a fox—singed and bloody as if stepping out of a furnace. Placing his hand on the paper on which Hughes was writing and leaving a bloody print, the apparition told him he must stop; the dream compelled Hughes to switch to the study of anthropology and archaeology for its wealth of folklore.

In his chapter "Dreams That Have Changed the World," from *Our Dreaming Mind,* Robert L. Van De Castle provides compelling examples of famous figures who have been guided by their dreams. Robert Penn Warren, author of *All The King's Men,* writes Van De Castle, often dreamed in great detail the scenes of his novels, even down to the dialogue. Likewise, American general George S. Patton not only believed he had been a soldier in previous lifetimes, but regularly received battle plans in his nightly dreams—including a successful surprise attack on German troops during the Battle of the Bulge.

My own creativity has been richly nourished by the deep springs of the psychic imagination. From my dreams, for instance, I have uncovered a wealth of ideas and images that have infused their magic into my writing. I have stumbled upon open books in ancient libraries, encountered otherworldly beings, and worn beautiful fabrics and jewels from other eras. Even while writing this book, I had a dream of Hermes in his trickster incarnation so unsettling and real I half expected the mischievous spirit to erase my computer files when I awoke. In the same way that my dreams enliven and inspire, so observing the movement of the planets as they trace a pattern in the heavens speaks to an archaic, wiser part of myself. Consulting an astrologer, I receive a unique kind of guidance based on the stars that helps to orient me on my mythic life map. Likewise evocative images from the tarot—such as the ace of cups with the image of a chalice with a dove descending into it—imparts meaning to my soul in ways that words rarely can.

One of my favorite tarot cards is that of the magician. In the Waite deck that I use, a man in a white robe with a red mantle and a serpent wound around his waist stands before a square red table. On it are four symbols: a chalice, a pentacle, a sword, and a wand, each one respectively representing one of the four elements of water, earth, air, and fire that are used to divide the tarot into four suits. Over the man's head hangs a figure eight lying on its side, a symbol of eternity. One arm lifts an illuminated candle skyward, while the other points down, imaging the ancient astrological principle "as above, so below."

This is the card of Hermes the magician; its symbols represents the transformative powers within our own psyches. As guide of souls, Hermes leads us to that part of ourselves that has foresight into the future, insight into the past—and the power to change the present. For without change, without genuine transformation, there is no real magic.

And how does Hermes as "inner magician" change the present? Through his capacity to "draw down" a different kind of energy that will lead to a change in our minds. As all mystics have taught, our thoughts are living forces that shape our realities. The way we think forms the invisible infrastructure of our everyday lives. Our running internal narrative is a plastic substance that molds the moods of our days. Thus by harnessing the communication god's power of transformation, we can literally change our lives by changing our minds.

A simple exercise shows how thought shapes physical reality. First, repeat the simple phrase "I can't do it. It won't work." After a while, your body and even the aura around you will bear the imprint of the energy patterns stirred by the atmosphere invoked by those words. Next, repeat the phrase "My life is a miracle of unfolding creativity." Like rays of sun beaming down on water creating a shimmering dance of light, the effect of these words is reanimating, stirring hope and joy to life in your soul. In the same way a small pebble cast into water spreads ripples outward, so the power of one small idea can create waves of energy in the inner world of our psyche.

This kind of mental magic is not a matter of simply glossing over negative thoughts with positive affirmations. As the word *transform* implies, the intent is literally to change the shape of our thoughts. We may, for instance, feel driven to stop trying something because we need to take a break from forcing something to happen a certain way.

But as nature abhors a vacuum, so our souls resist staying struck in one attitude and long for change and movement. Often, critical thoughts are the distorted forms of originally wholesome and life-oriented impulses. The self-defeating thought "I can't do it. It won't work," for instance, may have been a thought that first arose in response to our initial creative impulse being quashed by criticism or failure. As psychologist Allen Wheelis writes in *How People Change,* "Being the product of conditioning and being free to change do not war with each other. Both are true. They coexist, grow together in an upward spiral, and the growth of one furthers the growth of the other. The more . . . we prove ourselves to have been shaped by causes, the more opportunities we create for changing."

Thus mental transformation is a lot like restoring an old painting to its original beauty—scraping away the layers of accumulated color until the true picture emerges once again. By focusing our concentration on the golden glint of the positive shining through the dross of the negative, we can draw out the inherent potential for creative good within our minds that we were born with—but that got obscured along the way.

Indeed, regenerating tired and outworn thought patterns by rewriting and reimagining our mental scripts is a productive use of Mercury retrograde magic. Because this time period favors contemplation and meditation, another way to do this is through the repetition of a sacred phrase or prayer from a spiritual tradition. Or, we may purposefully screen our reading materials during this period, choosing to study only spiritual or inspirational texts. All incantation and ritual repetition, in fact, has at its heart the purpose of raising the energy by changing the mood or atmosphere that surrounds us.

Magic has always focused on both divining and altering the future. It was Odysseus's encounter in the Kingdom of the Dead with Tiresias the seer, for example, that allowed him to continue on his course with faith and courage. If Tiresias had not told the wandering king that he would ultimately return home to his kingdom, he might have given up too soon. In other words, Tiresias's prophecy allowed him to actually *see* for himself his way home through the shipwrecks and struggles that were still to come. At the same time, Odysseus's irregular, circular journey in which he was blown off course into encounters with fantastical creatures and underworld spirits benefited the king in another way. For if Odysseus had arrived home as quickly as he had originally intended, would he have valued his wife and homeland in the same profound way that he did after almost losing them entirely? In other words, by allowing us to step outside the box of our usual habits and routines, do our digressions, detours, and retrogrades actually serve to keep us on our true course? Just as the couple that breaks apart comes back together with a renewed sense of love and commitment, so losing and finding something helps us discern what is true and lasting from what is fleeting and impermanent.

This is the question that we face at the midpoint-phase of Mercury retrograde. Like Odysseus after his consultation with the seer Tiresias, we may have a glimpse of our future and a sense of what lies ahead. But though we have faith that there is an end in sight to our meandering journey, we are not quite there yet, and face more hurdles and obstructions on our way home. Guided by Hermes, however, we now have magic at our fingertips; with the help of our imagination, we can find a way to steer a course through the obstacles that still await us.

As we wait for our final approach to shore, we can occupy ourselves by writing down the insights and illuminations we have gained thus far. For like Odysseus, we have many tales to tell of our inner odyssey. Having grappled with our doubts and anxieties we are, perhaps, more firm in our resolve, more sure of where we are going and the value of what we already possess. Strengthened in who we really are, redirected to our true soul purpose, and rested in body, mind, and spirit, we enter the final leg of the long journey home.

10.

RETURN
TO THE WORLD

They came crowding out of their quarters, torch in
hand, flung their arms around Odysseus, hugging him,
home at last, and kissed his head and shoulders, seized
his hands, and he, overcome by a lovely longing, broke
down and wept . . .

—*The Odyssey,* translation by Robert Fagles

As *The Odyssey* draws to its thunderclap climax, Odysseus, the homesick traveler and weary warrior, finally reaches Ithaca. Again and again he has been blown off course, threatened by superhuman obstacles, and tempted with offers to forget his home and former identity. Yet always he finds a way out of the difficulties that waylay him, persisting in his quest. In the end, he arrives alone, having lost his ship and crew, all drowned in a hellish storm sent by the gods. Only Odysseus survives by clinging to the remnant of his ship's keel. Forewarned of the suitors who have overtaken his castle and courted his faithful wife, Penelope, in his absence, Odysseus shows up on the doorstep of his old home disguised as a bedraggled beggar. For before he can reclaim his kingdom, he must face his last hurdle—testing his wife's fidelity and winning back his castle from the usurpers.

While the great journey epic ultimately concludes with the loving reunion of husband and wife and the restoration of Odysseus's rulership, it is preceded by a period of suspense, palace intrigue, and a raging battle. Even after Odysseus has finally killed or cast out the suitors and shown his true identity to his friends and family, yet another battle threatens to erupt. At this point in the drama, Athena appears in her full glory, commanding Odysseus with these ringing words: "Royal son of Laertes, Odysseus master of exploits, hold back now! Call a halt to the great leveler, War! . . . So she commanded. He obeyed her, glad at heart. And Athena handed down her pacts of peace between both sides for all the years to come . . ."

With its cliff-hanging atmosphere of anxious uncertainty before the final resolution, the ending of *The Odyssey* vividly describes the return to life after a period of retreat away from the world. Often, discord arises, as the personal changes that have occurred and the new insights we have gained may cause friction as we try to squeeze ourselves back into former roles. Indeed, when Odysseus finally reveals himself to his son, so much more handsome does he appear that his son mistakes him for a god. This is symbolic, I think, of the long, slow transformation undertaken by Odysseus during his long sojourn; while he is still the same person, the challenges he has surmounted over the years have served to bring out the gold of his true being. He is larger than life because he has been forced to bring out the best in himself—or die or fail. At the same time, Odysseus's wife, Penelope, greets the husband to whom she has remained steadfastly loyal for over twenty years with skeptical wariness. Overcome at the news her husband has suddenly returned home, she is inexplicably seized with doubt when they see each other for the first time. "A long while she sat in silence . . . numbing wonder filled her heart as her eyes explored his face. One moment he seemed . . . Odysseus, to the life—the next, no, he was not the man she knew, a huddled mass of rags was all she saw." Only when Odysseus confirms his true identity to her by revealing the "secret signs" the couple once shared does Penelope surrender, dissolving into tears and throwing her arms around Odysseus. Passing these final challenges to his true identity, Odysseus is finally able to reclaim the kingdom he has nearly lost.

Though hard-won, his homecoming is rich with meaning and promise. For in the end, the battles Odysseus fought and the obstacles he overcame bring peace in their wake, as our own valiant inner struggles bring peace as well.

The elements at play during the final act of *The Odyssey* are often in force during the potent period of time when Mercury departs its retrograde retreat. This phase is called "stationary direct" and occurs when Mercury finally stops its backward motion, turns around, and begins to resume its forward journey along the zodiacal path. The planet does not do this quickly, however, but slowly, as if hovering in place—charging the atmosphere with a heightened tension. For three or four days before and after the day that Mercury begins to move forward, there is a collective in-drawing of breath like the pause after an orchestral crescendo; things hang in the balance, and we wait for the outcome of all our wanderings. The movie *Cold Mountain*—based on the Civil War novel by Charles Frazier that in turn reflects themes from *The Odyssey*—concludes on a similar note as the Greek legend. The wounded soldier Inman survives a perilous trek through the war-torn countryside to his hometown of Cold Mountain, North Carolina, where he is finally reunited with the woman whose love kept him alive. Yet upon his arrival, tired and suffering from his journey, he must fight one last battle with the local townsmen, who charge him as a deserter from the Confederate Army.

Astrologer Erin Sullivan describes this kind of period as the tense scene of a "threshold struggle." Apprehensive yet hopeful, we look back over the territory we have just traversed, assessing the new attitudes, ideas, intuitions, and plans that have begun to emerge out of the mists with clarity and shape, much as Odysseus espied the shoreline of Ithaca as he returned home from his long odyssey at sea.

Also like Odysseus or the soldier in *Cold Mountain,* however, we may face a final struggle as we stand at the threshold of achieving our dream: Once again, things may go "wrong" as they did at the outset of Mercury retrograde. Though we may think we know where we are headed, we may be in for even more surprises. We may experience the wind blowing us one way—when in fact we are being blown in just the opposite direction.

Often, in fact, the conclusion of this cycle brings unpredictable results— as if Mercury were a mischievous magician pulling one last trick out of his hat. The best way to be during this suspenseful several-day period, I have found, is to stay flexible and open-minded. Like a Zen master, we can rest in amused detachment as we watch the universe unfold along unforeseen lines. Poised but alert and aware, we can expect the unexpected. As the very nature of Mercury retrograde is to readjust that which has become outmoded and outgrown and to deepen our self-knowledge, it is a time to trust in the process that has been at work all along, reshaping our lives along new lines.

One of the best examples I have for illustrating this principle involves the very genesis of this book. I had just finished my book on women's spirituality, *Soul Sisters,* and was looking for a new project to work on. During a Mercury retrograde period, I had worked up several proposals to give to my editor and publisher for their consideration. Awaiting their response, I received a surprise e-mail from my editor—just as Mercury was about to station direct. "Would I like to write a book about Mercury retrograde?" he asked, knowing of my abiding interest in astrology. Of course, I would, accepting gladly this gift out of the blue while feeling awed all the while at the strange and magical workings of this little planet of destiny. It was an idea whose time had come, you could say— but one that I would never have seen coming myself. There is another Mercury retrograde lesson in this example, as well: Whatever we undertake during a retrograde period is likely to change form when the planet goes forward—just like my proposals, which I had crafted under a retrograde influence. This same principle can apply to our personal relationships. It is not unusual, for instance, for break-ups or arguments to occur around either the stationary retrograde or the stationary direct time periods. One woman I know resumed a former love affair at the outset of

Mercury retrograde. But as Mercury stationed direct, the man she was with—who had been involved with someone else while they were apart—suddenly changed his mind and returned to the other relationship. Because emotions can often run high at this time, some astrologers advise holding off on making any major decisions until Mercury has finally left the shadow period. Then, any new feeling or insights can be gradually integrated into our understanding of the relationship.

A very public and dramatic example of how Mercury can bring unintended results during its potentially volatile turnaround phase occurred during the 2000 presidential elections. Some months before, I had been sitting in a restaurant with a couple of friends, including astrologer Caroline Casey. Talking about the upcoming elections, she mused aloud about the fact that Mercury would be stationing direct the very night of the election and wondered about its possible implications. Other astrologers, as well, had foreseen possible disruptions.

As the whole world knows by now, the election results were thrown into complete confusion. "As the first vote was cast for president, Mercury was retrograde and just about to move back into the sign of Libra from Scorpio," wrote editor Tem Tarriktar in the February 2001 issue of the *Mountain Astrologer*. By the time the polls opened in Florida later that same day, he continued, "Mercury had backed into Libra . . . Mercury is retrograde during a U.S. presidential election about every 20 years. However, on *this* election night, at 9:24 P.M., EST, a few hours after the polls closed in Florida, Mercury *stopped and changed direction* from retrograde to direct motion, at 29.56 Libra . . ." Only minutes after Mercury turned around, it was announced that Florida was "too close to call." After midnight, the networks gave Florida to Bush, and Gore called to concede the election—only to call again and take it back. "By this time," wrote Tarriktar in his article, "astrologers everywhere must have been laughing—this was vintage Mercury retrograde mischief."

PLANTING AN OAR

All endings, astrology teaches, are but the beginnings of a new cycle. Thus the end of *The Odyssey* with its threshold struggle marks the onset of a new, more settled phase of life for the wandering hero, Odysseus. In his sojourn in the Kingdom of the Dead, Odysseus had been advised by the blind seer Tiresias to permanently anchor his homecoming by marching inland and firmly planting an oar in the soil. So, too, is it time to depart the mythic time zone of Mercury retrograde and ground our dreams in reality. As Mercury crosses the threshold and ventures out into clear horizons, so, too, can we begin to move forward. Now, we can forge ahead with our newly hatched plans, free to implement our creative inspirations in the crowded marketplace of the outer world or to commit to a new or existing relationship. Action takes precedence over thought, and we go about our work less hampered by unforeseen obstacles. It is a time to try out new things, to boldly experiment and break fresh ground. Having sorted through the moral dilemmas and personal decisions that we may have struggled with in the past several weeks, we may experience a peaceful clarity of mind and purpose. Those things that seemed doubtful and ambiguous now have an aura of resolution.

Often, issues that seemed weighty with significance suddenly evaporate, no longer as important as they once seemed—proving the "much ado about nothing" character that sometimes marks Mercury retrograde. Rebalanced, refocused, refreshed, and restored, it is like spring breaking out all over. And, in fact, the cycles of the natural world are reflected in the cycles of Mercury; while it is moving backward, it is as if all our energy goes into a wintry underground. The fields of our lives lie fallow, while deep in the dark earth new life germinates. When Mercury goes forward, however, these seeds shoot up through the soil, bursting out with color and the green, vital forces of life.

More important, however, Mercury retrograde returns to us our lost connection to soul. We rediscover, as the poet Matthew Arnold wrote, that "Strong is the soul, and wise, and beautiful: The seeds of godlike power are in us still: Gods are we, bards, saints and heroes, if we will." By slowing down and working with the wisdom of time, we overcome the modern-day addiction to speed, haste, and hurry. Instead, we dip into the well of timelessness that is the reservoir of mystics. We learn the patience of saints, for, as the ancient alchemical saying goes, "In your patience is your soul." Patience becomes the province of heroes as they grow old and age, too, as we saw with Odysseus. Had not Odysseus maintained steadfast patience and hope throughout the long years of his journey homeward, he would never have set eyes again on his wife and family. But Odysseus had help along the way from Hermes, the guide of souls. So, too, does the cycle of Mercury retrograde reconnect us to our own inner guide, the invisible one whose presence accompanies us on our journey through life.

Indeed, as disorienting and confusing as the jolts and shocks delivered by Mercury can sometimes be, they contain an important spiritual teaching: We cannot know the reasons why things happen the way they do. But if we can still our busy minds long enough, we may be able to connect to the ever-present spirit of guidance that can help deepen our understanding of life's mysteries. The story of Khidr and Moses helps to illustrate this principle. A Hermeslike figure found in the Koran and throughout Sufi lore, Khidr is the archetypal guide of souls. According to tradition, Khidr was the teacher of Moses who prepared him for prophethood. In a story told by the Sufi teacher Hazrat Inayat Khan in *Tales,* Moses and Khidr were traveling together. Moses' first lesson from Khidr was to practice the discipline of silence, and to keep quiet under all circumstances during their journey.

Soon, they came upon a child drowning, with the mother sobbing aloud because she could not help. Unable to keep quiet, Moses begged his master to intervene and save the child's life. Khidr's only response was a solemn "Quiet!" Inside himself, Moses boiled with anger at his master's seeming cruelty. Traveling further, they came

APPENDIX A

A Glossary of Basic Astrological
Terms and Symbols

The astrological chart is based on a mandala of signs and symbols that represents a pictorial map of the cosmos at a particular moment in time. The basic chart (see p. 173) is composed of four different components: 1) The houses, the twelve pie-shaped areas of the chart circle that each govern certain spheres of life. 2) The planets, each of which represent different life energies, archetypes, or gods and goddesses. 3) The signs—each planet and house is "colored," or influenced, by one of the twelve astrological signs. 4) Finally, the different planets make mathematical "aspects" to each other. Like characters in the play of our life, they are either strangers to one another, harmonize well, or are in tense conflict.

ASTROLOGICAL SYMBOLS:
THE TWELVE SIGNS IN THEIR ELEMENTS
AND MODES OF EXPRESSION

Each of the twelve signs is modified by falling in one of the four **elements** and one of the three **modes.** Like an adjective that modifies a noun, each of the four elements and three modes represents a set of characteristics that further defines the nature of each sign. The **fire** element is enthusiastic, dynamic, and passionate. The **air** element is objective, thoughtful, and detached. The **earth** element is stable, practical, and dependable. The **water** element is sympathetic, sensitive, and emotionally receptive.

The three modes are: **cardinal,** or forward-moving and self-directed; **fixed,** or stubborn and resistant to change; and **mutable,** or adaptable and changeable.

Aries ♈—A cardinal fire sign, Aries the ram is impulsive, pioneering, energetic, enterprising, and risk-taking.

Taurus ♉—A fixed earth sign, the sign of the bull is steady going, pleasure loving, stubborn, and dependable.

Gemini ♊—A mutable air sign, Gemini is intelligent, quick-witted, versatile, talkative, and perpetually curious.

Cancer ♋—A cardinal water sign, Cancer is emotional, sensitive, home loving, protective, maternal, nurturing.

Leo ♌—A fixed fire sign, Leo is courageous, creative, proud, dynamic, powerful, magnanimous, and warmhearted.

Virgo ♍—A mutable earth sign, Virgo is analytical, critical, service oriented, organized, and values integrity.

Libra ♎—A cardinal air sign, Libra is harmonious, artistic, loves beauty, and is relationship oriented.

Scorpio ♏—A fixed water sign, Scorpio is intense, secretive, passionate, occult, and loves mystery and depth.

Sagittarius ♐—A mutable fire sign, Sagittarius is philosophical, jovial, adventuresome, and broad-minded.

Capricorn ♑—A cardinal earth sign, Capricorn is practical, ambitious, cautious, responsible, driven, and achievement and success oriented.

Aquarius ♒—A fixed air sign, Aquarius is humanitarian, iconoclastic, intellectual, freedom loving, rebellious, and inventive.

Pisces ♓—A mutable water sign, Pisces is sensitive, mystical, artistic, intuitive, self-sacrificing, empathic, and imaginative.

THE PLANETS AND THE NODES

Sun ☉—Ruler of the sign of Leo, the Sun connotes will, individuality, spirit, conscious awareness, and one's core identity.

Moon ☽—Ruler of the sign of Cancer, the Moon represents personality, instinct, the unconscious, ancestral inheritance, and karma.

Mercury ☿—Ruler of the signs of Gemini and Virgo, Mercury symbolizes the mind, communication, and the link between spirit and matter.

Venus ♀—Ruler of the sign of Taurus and Libra, Venus represents Flove, affection, beauty, principles of fairness, and giving and receiving pleasure.

Mars ♂—Ruler of the sign of Aries, Mars symbolizes energy, initiative, courage, drive, friction, and combat.

Jupiter ♃—Ruler of the sign of Sagittarius, Jupiter represents faith, generosity, belief, self-confidence, and the principle of expansion.

Saturn ♄ —Ruler of the sign of Capricorn, Saturn connotes discipline, structure, limitation, duty and the principle of contraction.

Uranus ♅—Ruler of the sign of Aquarius, Uranus symbolizes independence, originality, and the forces of freedom and revolution.

Neptune ♆—Ruler of the sign of Pisces, Neptune signifies the qualities of both compassion and spirituality, as well as enchantment, illusion, and even delusion and confusion.

Pluto ♀—Ruler of the sign of Scorpio, Pluto represents regeneration, power, death and rebirth, and the deep collective unconscious.

North Node ☊—Wherever this symbol falls in the chart indicates an area of life that our soul has come to explore—our calling or life purpose.

South Node ☋—Wherever this symbol falls in the chart indicates an area of life that represents our inborn talents, inherited karma from past lifetimes, and the place where we can become stuck in old behaviors and ingrained habits.

THE HOUSES OF THE CHART

A Sample Chart

Ascendant/First House: The ascendant, or rising sign, is the sign on the horizon at birth. It marks the beginning of the first house and represents the point where we intersect with others and the persona we present to the world. The ascendant and first house also rule physical appearance, the self, and our individual disposition and approach to life. It is ruled by the sign of Aries and the planet Mars.

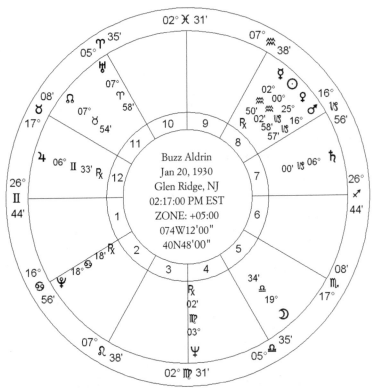

02° ♓ 31'

05° ♈ 35'

07° ♒ 38'

08° ♌
17° ♉

07° ♈ 58'

07° ♉ 54'

02° ♒ 00°
50' ♒ 25°
02' ♑ 16°
58' ♑
57' ♑

☿ ☉ ♀ 16° ♑ 56'
♂

06° ♊ 33' ♃ ℞

00' ♑ 06° ♄

26° ♊ 44'

26° ♐ 44'

Buzz Aldrin
Jan 20, 1930
Glen Ridge, NJ
02:17:00 PM EST
ZONE: +05:00
074W12'00"
40N48'00"

18' ♋ ℞
18° ♋

16° ♋ 56' ♇

08° ♏ 17'

34' ♎ 19° ☽

07° ♌ 38'

℞ 02' ♍ 03° ♇

05° ♎ 35'

02° ♍ 31'

DTP Standard

Second House: This sector in the chart governs money, material possessions, self-worth, and our sense of values and ethics. It is ruled by the sign of Taurus and the planet Venus.

Third House: This sector in the chart governs communications, writing and teaching, short trips, early education, and siblings. It is ruled by the sign of Gemini and the planet Mercury.

Nadir/Fourth House: The nadir marks the cusp of the fourth house, and is found at the bottom of the chart. This sector represents the foundation, the deepest inner base of oneself, and the foundations of family, home, and ancestral heritage. It is ruled by the sign of Cancer and the Moon.

Fifth House: This sector of the chart governs creativity, children, risk taking, pleasure, entertainment, and romance. It is ruled by the sign of Leo and the Sun.

Sixth House: This sector of the chart governs health, everyday rituals and routines, the work environment, pets, and service. It is ruled by the sign of Virgo and the planet Mercury.

Descendant/Seventh House: The descendant falls exactly opposite the ascendant, and marks the cusp of the seventh house. This segment in the chart represents partnerships, whether marriage or business. It is the area of the chart where we find ourselves in relationship with a significant "other." It is ruled by the sign of Libra and the planet Venus.

Eighth House: This sector of the chart represents other people's money, death and regeneration, occult mysteries, the personal unconscious, sexuality, and psychology. It is ruled by the sign of Scorpio and the planet Pluto.

Ninth House: This sector of the chart represents higher education, philosophy, religion, spirituality, foreign countries, and long journeys. It is ruled by the sign of Sagittarius and the planet Jupiter.

Midheaven/Tenth House: The midheaven falls exactly opposite the fourth house, and marks the cusp of the tenth house at the top of the chart. It is the sector that represents our career and standing in the world, the social world, and how we wish to be seen. It is ruled by the sign of Capricorn and the planet Saturn.

Eleventh House: This sector of the chart represents ideals and visions, the future, friendships, humanitarian causes, politics, and large organizations. It is ruled by the sign of Aquarius and the planet Uranus.

Twelfth House: The last house in the chart, this sector represents the collective unconscious, psychic phenomena, union with God, the mysterious and the hidden, death and reincarnation. It is ruled by the sign of Pisces and the planet Neptune.

THE MAJOR ASPECTS

These symbols represent the relationships that are formed between the different planets. They are based on the mathematical angles between two planets at a specific time.

The Conjunction ☌—An aspect that is formed when two planets fall in the same sign within 0 to 10 degrees. It signifies power, intensity, new beginnings, and the merger of two planetary principles.

The Trine △—An aspect of ease and harmony, the trine is symbolic of an inborn gift or genius. It is formed by a 120-degree angle.

The Square □—An aspect of conflict and challenge, the square represents difficulties and obstacles, often arising from divided loyalties within oneself. It is formed by a 90-degree angle.

The Opposition ☍—An aspect of tension that results from confrontation with an opposing force, the opposition can bring awareness, openness, and consciousness. It is formed by a 180-degree angle.

The Sextile ⚹—An aspect that signifies opportunity, the sextile represents inborn talents that need to be cultivated in order to flourish. It is formed by a 60-degree angle.

APPENDIX B

Recommended Resources on Mercury
and Basic Astrology

Athanassakis, A. N., trans. *The Homeric Hymns.* Baltimore and London: Johns Hopkins University Press, 1976. Translations of the ancient Greek hymns that depict the exploits and characteristics of the gods and goddesses of the Greek world, including Hermes.

Blaze, Chrissie. *Mercury Retrograde: Your Survival Guide to Astrology's Most Precarious Time of Year!* New York: Warner Books, 2002. Includes fascinating case studies, a comprehensive ephemeris of Mercury retrograde through the signs, and shows how to make the best use of this time period.

Edis, Freda. *The God Between: A Study of Astrological Mercury.* New York: Penguin/Arkana, 1995. An excellent in-depth book that covers both the mythological and astrological dimensions of Mercury as god and planet.

Grasse, Ray. *Signs of the Times: Unlocking the Symbolic Language of World Events.* Charlottesville, Va.: Hampton Roads, 2002. A "behind-the-scenes" look at the planetary alignments influencing some of the significant events of the twentieth and twenty-first centuries.

Greene, Liz, and Howard Sasportas. *The Inner Planets: Building Blocks of Personal Destiny.* York Beach, Me.: Samuel Weiser, 1993. A description of the psychological dynamics of each of the inner planets—Mercury, Venus, and Mars.

———. *The Luminaries: The Psychology of the Sun and the Moon in the Horoscope.* York Beach, Me.: Samuel Weiser, 1992. An imaginative and informative text on the archetypal function of the Sun and the Moon in one's chart.

Guttman, Ariel. *Mythic Astrology: Archetypal Powers in the Horoscope.* St. Paul, Minn.: Lewellyn Publications, 1998. An inspiring, insightful, and beautifully illustrated book on the myths and archetypes associated with the planets.

Hand, Robert. *Planets in Transit.* West Chester, Pa.: Para Research, 1976. A classic, basic text on the meaning of the different transits.

Hillman, Laurence, and Donna Spencer. *Alignments: How to Live in Harmony with the Universe.* New York: Lantern Books, 2002. Refreshingly contemporary and clearly written, this is an in-depth exploration of how the nodes reveal the soul's purpose and our calling in life.

Kerenyi, Karl. *Hermes: Guide of Souls.* Dallas: Spring Publications, 3rd printing 1987. A beautifully written contemplation of the many facets of the Greek god Hermes by a renowned mythographer and classical scholar.

Lundsted, Betty. *Astrological Insights into Personality.* San Diego: ACS Publications, 1980. A thoughtful explanation of the different aspects, with a psychological interpretation.

Michelson, Neil, and Rique Pottenger. *The American Ephemeris for the 21st Century 2000 to 2050 at Noon.* San Diego: ACS Publications, 1996. A convenient table of the daily movements of the Sun, Moon, and planets throughout the first half of the twentieth century.

Otto, Walter F. *The Homeric Gods: The Spiritual Significance of Greek Religion.* London and New York: Thames and Hudson/ Pantheon Books, 1979. A scholarly and illuminating examination of the archetypal significance of the Greek gods and goddesses, including Hermes.

Para Research, with preface by Robert Hand. *World Ephemeris for the 20th Century 1900 to 2000 at Noon.* Atglen, Pa.: Whitford Press, 1983. A convenient table of the daily movements of the Sun, Moon, and planets throughout the decades of the twentieth century.

Pelletier, Robert. *Planets in Aspect: Understanding Your Inner Dynamics.* Gloucester, Mass.: Para Research, 1974. A classic "cookbook" text that explains the meaning of each of the major astrological aspects.

Sasportas, Howard. *The Gods of Change: Pain, Crisis and the Transits of Uranus, Neptune and Pluto.* New York and London: Arkana/ Penguin, 1989. An in-depth look at the effects of outer planets' transits on the chart.

———. *The Twelve Houses: Understanding the Importance of the Houses in Your Astrological Birth Chart.* London: Thorsons/ HarperCollins, 1985. A definitive text that explains the areas of life ruled by each of the twelve houses of the birthchart.

Stein, Murray. *In Midlife.* Dallas: Spring Publications, 1983. A fascinating book that examines Hermes as the god of the midlife transition from a psychological and Jungian perspective.

Sullivan, Erin. *Retrograde Planets: Traversing the Inner Landscape.* York Beach, Me.: Samuel Weiser, 1992. A classic and superb text on the mythological and astrological phenomenon of the retrograde motion of all the planets, including how they operate in individual charts.

Whitfield, Peter. *Astrology: A History.* New York: Harry N. Abrams, 2001. Lavishly illustrated, this is a fascinating and compelling time line of the story of astrology through the ages.

Wilkinson, Robert. *A New Look at Mercury Retrograde.* York Beach, Me.: Samuel Weiser, 1997. A guide to the effects of Mercury retrograde, including a chart of historic events that occurred during Mercury retrograde.

ASTROLOGERS AND WEB SITES

To contact some of the astrologers quoted in this book for chart readings, here is a list of their e-mail addresses and Web sites.

Dana Gerhardt: dana@mooncircles.com; www.mooncircles.com
Ray Grasse: jupiter.enteract@rcn.com
Edith Hathaway: edithhathaway@shaw.ca
Laurence Hillman: laurence@lhillman.com; www.hillman.com
Lynn Koiner: koiner@starpower.net
Robert Schmidt: phaser@mindspring.com; www.projecthindsight.com
Erin Sullivan: erinsullivan@cybermesa.com; www.erinsullivan.com
Chakrapani Ullal: horastra@spiritmail.com;www.vedicastrology.com
April Kent: april@bigskyastrology.com;www.bigskyastrology.com

ASTROLOGY WEB SITES

www.alabe.com
This site offers a host of riches: free charts, personalized reports, articles for both beginning and advance astrologers, and software.

www.astrodienst.com
A good source for free charts. Also offers interpretations and other services.

www.astrologer.com
The Web site for Jungian analyst and astrologer Liz Greene. It of-

fers a variety of personalized reports by Greene and informative articles.

www.astrologyetal.com
A great site that offers hard-to-find astrology books and other astrology services.

www.astrologyzone.com
The Web site for the widely known astrologer Susan Miller. This delightful and easy-to-navigate site offers essays on Mercury retrograde, eclipse cycles, and other astrological events.

www.isisinstitute.com
A valuable resource for ordering tapes of lectures by such noted astrologers and thinkers as James Hillman and Laurence Hillman, Charles and Suzy Harvey, Richard Tarnas, and many others.

www.mountainastrologer.com
The Web site of the well-regarded magazine by the same name.

www.spiritualintrigue.com
If you want to tap into Mercury's trickster aspect, this is the site to visit. Hosted by visionary activist and astrologer Caroline Casey, this site features a wealth of articles and Casey's radio interviews with prominent thinkers.

www.stariq.com
A Web site by long-time astrologers Jeff Jawyer and Rick Levine. Excellent articles on current events and general astrological information.

APPENDIX C

Mercury Ephemeris 1910–2025

The following tables show Mercury's sign position through the months and years, as well as whether it is retrograde or direct. The dates are listed when Mercury changes sign, as well as when it shifts from direct motion to retrograde, or from retrograde to direct—these are the "stationary" points in the retrograde cycle and are marked with an *S*. For example, between March 12 and March 28, 1910, Mercury was direct in the sign of Pisces. So if you were born on March 14, 1910, your Mercury would be in Pisces and it would be direct. To illustrate Mercury's shift in motion, on May 14, 1910, Mercury turned retrograde in the sign of Gemini. On June 2, 1910, still moving retrograde, it backed up into the sign of Taurus, and on June 7 it turned direct while in the sign of Taurus.

An important note: This chart has been professionally programmed by Rique Pottenger of Astro Communications Services (ACS) and has been calculated for the Eastern Standard Time Zone. Because of the variations in time zones and times of day across the globe, if your birthday falls on one of the days when Mercury changes signs, it is possible that it could be in either of the two signs.

For instance, if your birthday is March 12, 1910, your Mercury could be in either Pisces or Aquarius. Only a chart drawn up for your date and time of birth will be able to show exactly in what sign your Mercury falls. Also, if your birthday falls either on, before, or after the day Mercury is listed as turning retrograde or direct, it is called "stationary" retrograde or "stationary" direct. This means that the importance of Mercury in your chart is even further intensified.

MERCURY SIGN CHANGES AND STATIONS
(EASTERN STANDARD TIME ZONE)

DATE	MOTION	SIGN
1910		
Jan. 4	Direct	Aquarius
Jan. 17	Retrograde S	Aquarius
Jan. 31	Retrograde	Capricorn
Feb. 7	Direct S	Capricorn
Feb. 15	Direct	Aquarius
Mar. 12	Direct	Pisces
Mar. 29	Direct	Aries
Apr. 13	Direct	Taurus
Apr. 30	Direct	Gemini
May 14	Retrograde S	Gemini
June 2	Retrograde	Taurus
June 7	Direct S	Taurus
June 12	Direct	Gemini
July 7	Direct	Cancer
July 21	Direct	Leo
Aug. 6	Direct	Virgo
Aug. 27	Direct	Libra
Sept. 13	Retrograde S	Libra
Sept. 28	Retrograde	Virgo
Oct. 5	Direct S	Virgo
Oct. 12	Direct	Libra
Oct. 31	Direct	Scorpio

| Nov. 19 | Direct | Sagittarius |
| Dec. 8 | Direct | Capricorn |

1911

Jan. 1	Retrograde S	Capricorn
Jan. 21	Direct S	Capricorn
Feb. 13	Direct	Aquarius
Mar. 5	Direct	Pisces
Mar. 21	Direct	Aries
Apr. 5	Direct	Taurus
Apr. 25	Retrograde S	Taurus
May 18	Direct S	Taurus
June 13	Direct	Gemini
June 29	Direct	Cancer
July 13	Direct	Leo
July 30	Direct	Virgo
Aug. 27	Retrograde S	Virgo
Sept. 19	Direct S	Virgo
Oct. 7	Direct	Libra
Oct. 24	Direct	Scorpio
Nov. 12	Direct	Sagittarius
Dec. 3	Direct	Capricorn
Dec. 16	Retrograde S	Capricorn
Dec. 27	Retrograde	Sagittarius

1912

Jan. 5	Direct S	Sagittarius
Jan. 15	Direct	Capricorn
Feb. 7	Direct	Aquarius
Feb. 25	Direct	Pisces
Mar. 12	Direct	Aries
Apr. 5	Retrograde S	Aries
Apr. 29	Direct S	Aries
May 17	Direct	Taurus
June 5	Direct	Gemini
June 19	Direct	Cancer
July 4	Direct	Leo

July 26	Direct	Virgo
Aug. 8	Retrograde S	Virgo
Aug. 21	Retrograde	Leo
Sept. 1	Direct S	Leo
Sept. 10	Direct	Virgo
Sept. 28	Direct	Libra
Oct. 15	Direct	Scorpio
Nov. 4	Direct	Sagittarius
Nov. 29	Retrograde S	Sagittarius
Dec. 19	Direct S	Sagittarius

1913

Jan. 10	Direct	Capricorn
Jan. 30	Direct	Aquarius
Feb. 16	Direct	Pisces
Mar. 5	Direct	Aries
Mar. 18	Retrograde S	Aries
Apr. 7	Retrograde	Pisces
Apr. 11	Direct S	Pisces
Apr. 14	Direct	Aries
May 12	Direct	Taurus
May 28	Direct	Gemini
June 11	Direct	Cancer
June 28	Direct	Leo
July 21	Retrograde S	Leo
Aug. 14	Direct S	Leo
Sept. 4	Direct	Virgo
Sept. 20	Direct	Libra
Oct. 8	Direct	Scorpio
Oct. 30	Direct	Sagittarius
Nov. 13	Retrograde S	Sagittarius
Nov. 23	Retrograde	Scorpio
Dec. 2	Direct S	Scorpio
Dec. 13	Direct	Sagittarius

1914

Jan. 4	Direct	Capricorn
Jan. 22	Direct	Aquarius

Feb. 9	Direct	Pisces
Mar. 1	Retrograde S	Pisces
Mar. 24	Direct S	Pisces
Apr. 16	Direct	Aries
May 5	Direct	Taurus
May 19	Direct	Gemini
June 3	Direct	Cancer
July 3	Retrograde S	Cancer
July 27	Direct S	Cancer
Aug. 11	Direct	Leo
Aug. 27	Direct	Virgo
Sept. 12	Direct	Libra
Oct. 2	Direct	Scorpio
Oct. 27	Retrograde S	Scorpio
Nov. 16	Direct S	Scorpio
Dec. 8	Direct	Sagittarius
Dec. 27	Direct	Capricorn

1915

Jan. 15	Direct	Aquarius
Feb. 2	Direct	Pisces
Feb. 12	Retrograde S	Pisces
Feb. 23	Retrograde	Aquarius
Mar. 6	Direct S	Aquarius
Mar. 19	Direct	Pisces
Apr. 11	Direct	Aries
Apr. 27	Direct	Taurus
May 11	Direct	Gemini
May 29	Direct	Cancer
June 14	Retrograde S	Cancer
July 8	Direct S	Cancer
Aug. 4	Direct	Leo
Aug. 19	Direct	Virgo
Sept. 5	Direct	Libra
Sept. 28	Direct	Scorpio
Oct. 10	Retrograde S	Scorpio
Oct. 21	Retrograde	Libra
Oct. 31	Direct S	Libra
Nov. 11	Direct	Scorpio

| Dec. 1 | Direct | Sagittarius |
| Dec. 20 | Direct | Capricorn |

1916

Jan. 8	Direct	Aquarius
Jan. 27	Retrograde S	Aquarius
Feb. 17	Direct S	Aquarius
Mar. 15	Direct	Pisces
Apr. 2	Direct	Aries
Apr. 17	Direct	Taurus
May 2	Direct	Gemini
May 25	Retrograde S	Gemini
June 18	Direct S	Gemini
July 10	Direct	Cancer
July 26	Direct	Leo
Aug. 10	Direct	Virgo
Aug. 29	Direct	Libra
Sept. 23	Retrograde S	Libra
Oct. 14	Direct S	Libra
Nov. 4	Direct	Scorpio
Nov. 23	Direct	Sagittarius
Dec. 12	Direct	Capricorn

1917

Jan. 1	Direct	Aquarius
Jan. 10	Retrograde S	Aquarius
Jan. 18	Retrograde	Capricorn
Jan. 31	Direct S	Capricorn
Feb. 15	Direct	Aquarius
Mar. 8	Direct	Pisces
Mar. 25	Direct	Aries
Apr. 9	Direct	Taurus
May 5	Retrograde S	Taurus
May 29	Direct S	Taurus
June 14	Direct	Gemini
July 3	Direct	Cancer
July 17	Direct	Leo
Aug. 3	Direct	Virgo

Aug. 27	Direct	Libra
Sept. 5	Retrograde S	Libra
Sept. 14	Retrograde	Virgo
Sept. 28	Direct S	Virgo
Oct. 10	Direct	Libra
Oct. 28	Direct	Scorpio
Nov. 16	Direct	Sagittarius
Dec. 5	Direct	Capricorn
Dec. 25	Retrograde S	Capricorn

1918

Jan. 14	Direct S	Capricorn
Feb. 10	Direct	Aquarius
Mar. 1	Direct	Pisces
Mar. 17	Direct	Aries
Apr. 2	Direct	Taurus
Apr. 16	Retrograde S	Taurus
May 10	Direct S	Taurus
June 10	Direct	Gemini
June 25	Direct	Cancer
July 9	Direct	Leo
July 28	Direct	Virgo
Aug. 19	Retrograde S	Virgo
Sept. 11	Direct S	Virgo
Oct. 3	Direct	Libra
Oct. 20	Direct	Scorpio
Nov. 9	Direct	Sagittarius
Dec. 1	Direct	Capricorn
Dec. 9	Retrograde S	Capricorn
Dec. 15	Retrograde	Sagittarius
Dec. 29	Direct S	Sagittarius

1919

Jan. 13	Direct	Capricorn
Feb. 4	Direct	Aquarius
Feb. 21	Direct	Pisces
Mar. 9	Direct	Aries
Mar. 29	Retrograde S	Aries

Apr. 22	Direct S	Aries
May 16	Direct	Taurus
June 2	Direct	Gemini
June 16	Direct	Cancer
July 2	Direct	Leo
Aug. 1	Retrograde S	Leo
Aug. 25	Direct S	Leo
Sept. 9	Direct	Virgo
Sept. 25	Direct	Libra
Oct. 13	Direct	Scorpio
Nov. 3	Direct	Sagittarius
Nov. 23	Retrograde S	Sagittarius
Dec. 12	Direct S	Sagittarius

1920

Jan. 8	Direct	Capricorn
Jan. 27	Direct	Aquarius
Feb. 13	Direct	Pisces
Mar. 3	Direct	Aries
Mar. 11	Retrograde S	Aries
Mar. 19	Retrograde	Pisces
Apr. 3	Direct S	Pisces
Apr. 17	Direct	Aries
May 9	Direct	Taurus
May 24	Direct	Gemini
June 7	Direct	Cancer
June 26	Direct	Leo
July 13	Retrograde S	Leo
Aug. 3	Retrograde	Cancer
Aug. 6	Direct S	Cancer
Aug. 10	Direct	Leo
Aug. 31	Direct	Virgo
Sept. 16	Direct	Libra
Oct. 5	Direct	Scorpio
Oct. 30	Direct	Sagittarius
Nov. 5	Retrograde S	Sagittarius
Nov. 11	Retrograde	Scorpio
Nov. 25	Direct S	Scorpio

| Dec. 11 | Direct | Sagittarius |
| Dec. 31 | Direct | Capricorn |

1921

Jan. 19	Direct	Aquarius
Feb. 5	Direct	Pisces
Feb. 22	Retrograde S	Pisces
Mar. 16	Direct S	Pisces
Apr. 14	Direct	Aries
May 1	Direct	Taurus
May 15	Direct	Gemini
May 31	Direct	Cancer
June 25	Retrograde S	Cancer
July 19	Direct S	Cancer
Aug. 8	Direct	Leo
Aug. 23	Direct	Virgo
Sept. 9	Direct	Libra
Sept. 29	Direct	Scorpio
Oct. 20	Retrograde S	Scorpio
Nov. 9	Direct S	Scorpio
Dec. 5	Direct	Sagittarius
Dec. 24	Direct	Capricorn

1922

Jan. 11	Direct	Aquarius
Feb. 1	Direct	Pisces
Feb. 5	Retrograde S	Pisces
Feb. 9	Retrograde	Aquarius
Feb. 27	Direct S	Aquarius
Mar. 18	Direct	Pisces
Apr. 7	Direct	Aries
Apr. 23	Direct	Taurus
May 7	Direct	Gemini
June 1	Direct	Cancer
June 6	Retrograde S	Cancer
June 11	Retrograde	Gemini
June 30	Direct S	Gemini

July 14	Direct	Cancer
July 31	Direct	Leo
Aug. 15	Direct	Virgo
Sept. 2	Direct	Libra
Oct. 1	Direct	Scorpio
Oct. 3	Retrograde S	Scorpio
Oct. 5	Retrograde	Libra
Oct. 24	Direct S	Libra
Nov. 9	Direct	Scorpio
Nov. 28	Direct	Sagittarius
Dec. 17	Direct	Capricorn

1923

Jan. 5	Direct	Aquarius
Jan. 20	Retrograde S	Aquarius
Feb. 6	Retrograde	Capricorn
Feb. 10	Direct S	Capricorn
Feb. 14	Direct	Aquarius
Mar. 13	Direct	Pisces
Mar. 30	Direct	Aries
Apr. 14	Direct	Taurus
May 1	Direct	Gemini
May 17	Retrograde S	Gemini
June 10	Direct S	Gemini
July 8	Direct	Cancer
July 23	Direct	Leo
Aug. 7	Direct	Virgo
Aug. 28	Direct	Libra
Sept. 16	Retrograde S	Libra
Oct. 4	Retrograde	Virgo
Oct. 8	Direct S	Virgo
Oct. 12	Direct	Libra
Nov. 2	Direct	Scorpio
Nov. 20	Direct	Sagittarius
Dec. 10	Direct	Capricorn

1924

Jan. 4	Retrograde S	Capricorn
Jan. 24	Direct S	Capricorn
Feb. 14	Direct	Aquarius
Mar. 5	Direct	Pisces
Mar. 21	Direct	Aries
Apr. 5	Direct	Taurus
Apr. 27	Retrograde S	Taurus
May 21	Direct S	Taurus
June 13	Direct	Gemini
June 29	Direct	Cancer
July 13	Direct	Leo
July 30	Direct	Virgo
Aug. 29	Retrograde S	Virgo
Sept. 21	Direct S	Virgo
Oct. 7	Direct	Libra
Oct. 24	Direct	Scorpio
Nov. 12	Direct	Sagittarius
Dec. 3	Direct	Capricorn
Dec. 18	Retrograde S	Capricorn
Dec. 31	Retrograde	Sagittarius

1925

Jan. 7	Direct S	Sagittarius
Jan. 14	Direct	Capricorn
Feb. 7	Direct	Aquarius
Feb. 25	Direct	Pisces
Mar. 13	Direct	Aries
Apr. 1	Direct	Taurus
Apr. 8	Retrograde S	Taurus
Apr. 16	Retrograde	Aries
May 2	Direct S	Aries
May 17	Direct	Taurus
June 6	Direct	Gemini
June 21	Direct	Cancer
July 5	Direct	Leo
July 26	Direct	Virgo
Aug. 11	Retrograde S	Virgo

Aug. 27	Retrograde	Leo
Sept. 4	Direct S	Leo
Sept. 11	Direct	Virgo
Sept. 29	Direct	Libra
Oct. 17	Direct	Scorpio
Nov. 5	Direct	Sagittarius
Dec. 2	Retrograde S	Sagittarius
Dec. 21	Direct S	Sagittarius

1926

Jan. 11	Direct	Capricorn
Jan. 31	Direct	Aquarius
Feb. 18	Direct	Pisces
Mar. 6	Direct	Aries
Mar. 21	Retrograde S	Aries
Apr. 14	Direct S	Aries
May 13	Direct	Taurus
May 29	Direct	Gemini
June 12	Direct	Cancer
June 29	Direct	Leo
July 24	Retrograde S	Leo
Aug. 17	Direct S	Leo
Sept. 6	Direct	Virgo
Sept. 22	Direct	Libra
Oct. 10	Direct	Scorpio
Oct. 31	Direct	Sagittarius
Nov. 15	Retrograde S	Sagittarius
Nov. 28	Retrograde	Scorpio
Dec. 5	Direct S	Scorpio
Dec. 14	Direct	Sagittarius

1927

Jan. 5	Direct	Capricorn
Jan. 24	Direct	Aquarius
Feb. 10	Direct	Pisces
Mar. 4	Retrograde S	Pisces
Mar. 27	Direct S	Pisces
Apr. 17	Direct	Aries

May 6	Direct	Taurus
May 21	Direct	Gemini
June 4	Direct	Cancer
June 29	Direct	Leo
July 6	Retrograde S	Leo
July 14	Retrograde	Cancer
July 30	Direct S	Cancer
Aug. 12	Direct	Leo
Aug. 29	Direct	Virgo
Sept. 14	Direct	Libra
Oct. 3	Direct	Scorpio
Oct. 30	Retrograde S	Scorpio
Nov. 19	Direct S	Scorpio
Dec. 9	Direct	Sagittarius
Dec. 29	Direct	Capricorn

1928

Jan. 16	Direct	Aquarius
Feb. 3	Direct	Pisces
Feb. 15	Retrograde S	Pisces
Feb. 29	Retrograde	Aquarius
Mar. 8	Direct S	Aquarius
Mar. 18	Direct	Pisces
Apr. 11	Direct	Aries
Apr. 27	Direct	Taurus
May 11	Direct	Gemini
May 29	Direct	Cancer
June 16	Retrograde S	Cancer
July 11	Direct S	Cancer
Aug. 5	Direct	Leo
Aug. 19	Direct	Virgo
Sept. 5	Direct	Libra
Sept. 27	Direct	Scorpio
Oct. 12	Retrograde S	Scorpio
Oct. 25	Retrograde	Libra
Nov. 2	Direct S	Libra
Nov. 11	Direct	Scorpio
Dec. 1	Direct	Sagittarius
Dec. 21	Direct	Capricorn

1929

Jan. 8	Direct	Aquarius
Jan. 29	Retrograde S	Aquarius
Feb. 19	Direct S	Aquarius
Mar. 16	Direct	Pisces
Apr. 4	Direct	Aries
Apr. 19	Direct	Taurus
May 4	Direct	Gemini
May 28	Retrograde S	Gemini
June 21	Direct S	Gemini
July 12	Direct	Cancer
July 27	Direct	Leo
Aug. 11	Direct	Virgo
Aug. 30	Direct	Libra
Sept. 25	Retrograde S	Libra
Oct. 17	Direct S	Libra
Nov. 6	Direct	Scorpio
Nov. 24	Direct	Sagittarius
Dec. 13	Direct	Capricorn

1930

Jan. 2	Direct	Aquarius
Jan. 13	Retrograde S	Aquarius
Jan. 23	Retrograde	Capricorn
Feb. 2	Direct S	Capricorn
Feb. 15	Direct	Aquarius
Mar. 10	Direct	Pisces
Mar. 27	Direct	Aries
Apr. 10	Direct	Taurus
May 1	Direct	Gemini
May 9	Retrograde S	Gemini
May 17	Retrograde	Taurus
June 1	Direct S	Taurus
June 15	Direct	Gemini
July 5	Direct	Cancer
July 19	Direct	Leo
Aug. 4	Direct	Virgo
Aug. 26	Direct	Libra

Sept. 8	Retrograde S	Libra
Sept. 20	Retrograde	Virgo
Oct. 1	Direct S	Virgo
Oct. 11	Direct	Libra
Oct. 29	Direct	Scorpio
Nov. 17	Direct	Sagittarius
Dec. 7	Direct	Capricorn
Dec. 28	Retrograde S	Capricorn

1931

Jan. 17	Direct S	Capricorn
Feb. 11	Direct	Aquarius
Mar. 2	Direct	Pisces
Mar. 19	Direct	Aries
Apr. 3	Direct	Taurus
Apr. 20	Retrograde S	Taurus
May 13	Direct S	Taurus
June 11	Direct	Gemini
June 26	Direct	Cancer
July 11	Direct	Leo
July 29	Direct	Virgo
Aug. 22	Retrograde S	Virgo
Sept. 14	Direct S	Virgo
Oct. 4	Direct	Libra
Oct. 22	Direct	Scorpio
Nov. 10	Direct	Sagittarius
Dec. 2	Direct	Capricorn
Dec. 12	Retrograde S	Capricorn
Dec. 20	Retrograde	Sagittarius
Dec. 31	Direct S	Sagittarius

1932

Jan. 14	Direct	Capricorn
Feb. 5	Direct	Aquarius
Feb. 23	Direct	Pisces
Mar. 10	Direct	Aries
Mar. 31	Retrograde S	Aries
Apr. 24	Direct S	Aries

May 16	Direct	Taurus
June 3	Direct	Gemini
June 17	Direct	Cancer
July 2	Direct	Leo
July 28	Direct	Virgo
Aug. 3	Retrograde S	Virgo
Aug. 10	Retrograde	Leo
Aug. 27	Direct S	Leo
Sept. 9	Direct	Virgo
Sept. 26	Direct	Libra
Oct. 13	Direct	Scorpio
Nov. 3	Direct	Sagittarius
Nov. 24	Retrograde S	Sagittarius
Dec. 14	Direct S	Sagittarius

1933

Jan. 8	Direct	Capricorn
Jan. 28	Direct	Aquarius
Feb. 14	Direct	Pisces
Mar. 3	Direct	Aries
Mar. 14	Retrograde S	Aries
Mar. 26	Retrograde	Pisces
Apr. 6	Direct S	Pisces
Apr. 17	Direct	Aries
May 10	Direct	Taurus
May 25	Direct	Gemini
June 8	Direct	Cancer
June 27	Direct	Leo
July 16	Retrograde S	Leo
Aug. 10	Direct S	Leo
Sept. 2	Direct	Virgo
Sept. 18	Direct	Libra
Oct. 6	Direct	Scorpio
Oct. 30	Direct	Sagittarius
Nov. 8	Retrograde S	Sagittarius
Nov. 16	Retrograde	Scorpio
Nov. 28	Direct S	Scorpio
Dec. 12	Direct	Sagittarius

1934

Jan. 1	Direct	Capricorn
Jan. 20	Direct	Aquarius
Feb. 6	Direct	Pisces
Feb. 25	Retrograde S	Pisces
Mar. 19	Direct S	Pisces
Apr. 15	Direct	Aries
May 2	Direct	Taurus
May 17	Direct	Gemini
June 1	Direct	Cancer
June 28	Retrograde S	Cancer
July 22	Direct S	Cancer
Aug. 9	Direct	Leo
Aug. 25	Direct	Virgo
Sept. 10	Direct	Libra
Sept. 30	Direct	Scorpio
Oct. 23	Retrograde S	Scorpio
Nov. 12	Direct S	Scorpio
Dec. 6	Direct	Sagittarius
Dec. 25	Direct	Capricorn

1935

Jan. 13	Direct	Aquarius
Feb. 1	Direct	Pisces
Feb. 8	Retrograde S	Pisces
Feb. 15	Retrograde	Aquarius
Mar. 2	Direct S	Aquarius
Mar. 19	Direct	Pisces
Apr. 8	Direct	Aries
Apr. 24	Direct	Taurus
May 8	Direct	Gemini
May 30	Direct	Cancer
June 9	Retrograde S	Cancer
June 20	Retrograde	Gemini
July 3	Direct S	Gemini
July 14	Direct	Cancer
Aug. 2	Direct	Leo
Aug. 17	Direct	Virgo

Sept. 3	Direct	Libra
Sept. 28	Direct	Scorpio
Oct. 6	Retrograde S	Scorpio
Oct. 12	Retrograde	Libra
Oct. 27	Direct S	Libra
Nov. 10	Direct	Scorpio
Nov. 29	Direct	Sagittarius
Dec. 18	Direct	Capricorn

1936

Jan. 6	Direct	Aquarius
Jan. 23	Retrograde S	Aquarius
Feb. 13	Direct S	Aquarius
Mar. 13	Direct	Pisces
Mar. 31	Direct	Aries
Apr. 15	Direct	Taurus
May 1	Direct	Gemini
May 20	Retrograde S	Gemini
June 12	Direct S	Gemini
July 9	Direct	Cancer
July 23	Direct	Leo
Aug. 8	Direct	Virgo
Aug. 27	Direct	Libra
Sept. 18	Retrograde S	Libra
Oct. 10	Direct S	Libra
Nov. 2	Direct	Scorpio
Nov. 21	Direct	Sagittarius
Dec. 10	Direct	Capricorn

1937

Jan. 1	Direct	Aquarius
Jan. 6	Retrograde S	Aquarius
Jan. 10	Retrograde	Capricorn
Jan. 26	Direct S	Capricorn
Feb. 14	Direct	Aquarius
Mar. 6	Direct	Pisces
Mar. 23	Direct	Aries
Apr. 7	Direct	Taurus

Apr. 30	Retrograde S	Taurus
May 24	Direct S	Taurus
June 14	Direct	Gemini
July 1	Direct	Cancer
July 15	Direct	Leo
Aug. 1	Direct	Virgo
Sept. 1	Retrograde S	Virgo
Sept. 23	Direct S	Virgo
Oct. 8	Direct	Libra
Oct. 26	Direct	Scorpio
Nov. 14	Direct	Sagittarius
Dec. 4	Direct	Capricorn
Dec. 21	Retrograde S	Capricorn

1938

Jan. 7	Retrograde	Sagittarius
Jan. 10	Direct S	Sagittarius
Jan. 13	Direct	Capricorn
Feb. 8	Direct	Aquarius
Feb. 27	Direct	Pisces
Mar. 15	Direct	Aries
Apr. 1	Direct	Taurus
Apr. 11	Retrograde S	Taurus
Apr. 23	Retrograde	Aries
May 5	Direct S	Aries
May 16	Direct	Taurus
June 8	Direct	Gemini
June 22	Direct	Cancer
July 7	Direct	Leo
July 27	Direct	Virgo
Aug. 14	Retrograde S	Virgo
Sept. 3	Retrograde	Leo
Sept. 7	Direct S	Leo
Sept. 10	Direct	Virgo
Oct. 1	Direct	Libra
Oct. 18	Direct	Scorpio
Nov. 7	Direct	Sagittarius
Dec. 4	Retrograde S	Sagittarius
Dec. 24	Direct S	Sagittarius

1939

Jan. 12	Direct	Capricorn
Feb. 1	Direct	Aquarius
Feb. 19	Direct	Pisces
Mar. 7	Direct	Aries
Mar. 24	Retrograde S	Aries
Apr. 17	Direct S	Aries
May 14	Direct	Taurus
May 31	Direct	Gemini
June 14	Direct	Cancer
June 30	Direct	Leo
July 27	Retrograde S	Leo
Aug. 20	Direct S	Leo
Sept. 7	Direct	Virgo
Sept. 23	Direct	Libra
Oct. 11	Direct	Scorpio
Nov. 1	Direct	Sagittarius
Nov. 18	Retrograde S	Sagittarius
Dec. 3	Retrograde	Scorpio
Dec. 8	Direct S	Scorpio
Dec. 14	Direct	Sagittarius

1940

Jan. 6	Direct	Capricorn
Jan. 25	Direct	Aquarius
Feb. 11	Direct	Pisces
Mar. 4	Direct	Aries
Mar. 6	Retrograde S	Aries
Mar. 8	Retrograde	Pisces
Mar. 29	Direct S	Pisces
Apr. 17	Direct	Aries
May 7	Direct	Taurus
May 21	Direct	Gemini
June 5	Direct	Cancer
June 26	Direct	Leo
July 8	Retrograde S	Leo
July 21	Retrograde	Cancer
Aug. 1	Direct S	Cancer

Aug. 11	Direct	Leo
Aug. 29	Direct	Virgo
Sept. 14	Direct	Libra
Oct. 3	Direct	Scorpio
Nov. 1	Retrograde S	Scorpio
Nov. 21	Direct S	Scorpio
Dec. 9	Direct	Sagittarius
Dec. 29	Direct	Capricorn

1941

Jan. 17	Direct	Aquarius
Feb. 3	Direct	Pisces
Feb. 17	Retrograde S	Pisces
Mar. 7	Retrograde	Aquarius
Mar. 11	Direct S	Aquarius
Mar. 16	Direct	Pisces
Apr. 12	Direct	Aries
Apr. 29	Direct	Taurus
May 13	Direct	Gemini
May 29	Direct	Cancer
June 20	Retrograde S	Cancer
July 14	Direct S	Cancer
Aug. 6	Direct	Leo
Aug. 21	Direct	Virgo
Sept. 7	Direct	Libra
Sept. 28	Direct	Scorpio
Oct. 15	Retrograde S	Scorpio
Oct. 30	Retrograde	Libra
Nov. 5	Direct S	Libra
Nov. 12	Direct	Scorpio
Dec. 3	Direct	Sagittarius
Dec. 22	Direct	Capricorn

1942

Jan. 9	Direct	Aquarius
Feb. 1	Retrograde S	Aquarius
Feb. 22	Direct S	Aquarius
Mar. 17	Direct	Pisces

Apr. 5	Direct	Aries
Apr. 20	Direct	Taurus
May 5	Direct	Gemini
May 31	Retrograde S	Gemini
June 24	Direct S	Gemini
July 13	Direct	Cancer
July 29	Direct	Leo
Aug. 13	Direct	Virgo
Aug. 31	Direct	Libra
Sept. 28	Retrograde S	Libra
Oct. 20	Direct S	Libra
Nov. 7	Direct	Scorpio
Nov. 26	Direct	Sagittarius
Dec. 15	Direct	Capricorn

1943

Jan. 3	Direct	Aquarius
Jan. 16	Retrograde S	Aquarius
Jan. 28	Retrograde	Capricorn
Feb. 5	Direct S	Capricorn
Feb. 15	Direct	Aquarius
Mar. 11	Direct	Pisces
Mar. 28	Direct	Aries
Apr. 12	Direct	Taurus
Apr. 30	Direct	Gemini
May 12	Retrograde S	Gemini
May 26	Retrograde	Taurus
June 5	Direct S	Taurus
June 14	Direct	Gemini
July 6	Direct	Cancer
July 20	Direct	Leo
Aug. 5	Direct	Virgo
Aug. 27	Direct	Libra
Sept. 11	Retrograde S	Libra
Sept. 25	Retrograde	Virgo
Oct. 3	Direct S	Virgo
Oct. 12	Direct	Libra
Oct. 31	Direct	Scorpio
Nov. 18	Direct	Sagittarius

| Dec. 8 | Direct | Capricorn |
| Dec. 30 | Retrograde S | Capricorn |

1944

Jan. 20	Direct S	Capricorn
Feb. 12	Direct	Aquarius
Mar. 3	Direct	Pisces
Mar. 19	Direct	Aries
Apr. 3	Direct	Taurus
Apr. 22	Retrograde S	Taurus
May 16	Direct S	Taurus
June 11	Direct	Gemini
June 27	Direct	Cancer
July 11	Direct	Leo
July 29	Direct	Virgo
Aug. 24	Retrograde S	Virgo
Sept. 16	Direct S	Virgo
Oct. 5	Direct	Libra
Oct. 22	Direct	Scorpio
Nov. 10	Direct	Sagittarius
Dec. 1	Direct	Capricorn
Dec. 13	Retrograde S	Capricorn
Dec. 24	Retrograde	Sagittarius

1945

Jan. 2	Direct S	Sagittarius
Jan. 14	Direct	Capricorn
Feb. 5	Direct	Aquarius
Feb. 23	Direct	Pisces
Mar. 11	Direct	Aries
Apr. 3	Retrograde S	Aries
Apr. 27	Direct S	Aries
May 16	Direct	Taurus
June 4	Direct	Gemini
June 18	Direct	Cancer
July 3	Direct	Leo
July 26	Direct	Virgo
Aug. 6	Retrograde S	Virgo

Aug. 17	Retrograde	Leo
Aug. 30	Direct S	Leo
Sept. 10	Direct	Virgo
Sept. 27	Direct	Libra
Oct. 15	Direct	Scorpio
Nov. 4	Direct	Sagittarius
Nov. 27	Retrograde S	Sagittarius
Dec. 17	Direct S	Sagittarius

1946

Jan. 9	Direct	Capricorn
Jan. 29	Direct	Aquarius
Feb. 15	Direct	Pisces
Mar. 4	Direct	Aries
Mar. 16	Retrograde S	Aries
Apr. 1	Retrograde	Pisces
Apr. 9	Direct S	Pisces
Apr. 16	Direct	Aries
May 11	Direct	Taurus
May 27	Direct	Gemini
June 10	Direct	Cancer
June 28	Direct	Leo
July 19	Retrograde S	Leo
Aug. 13	Direct S	Leo
Sept. 3	Direct	Virgo
Sept. 19	Direct	Libra
Oct. 8	Direct	Scorpio
Oct. 30	Direct	Sagittarius
Nov. 11	Retrograde S	Sagittarius
Nov. 21	Retrograde	Scorpio
Dec. 1	Direct S	Scorpio
Dec. 13	Direct	Sagittarius

1947

Jan. 3	Direct	Capricorn
Jan. 22	Direct	Aquarius
Feb. 8	Direct	Pisces
Feb. 27	Retrograde S	Pisces

Mar. 22	Direct S	Pisces
Apr. 16	Direct	Aries
May 4	Direct	Taurus
May 18	Direct	Gemini
June 2	Direct	Cancer
July 1	Retrograde S	Cancer
July 25	Direct S	Cancer
Aug. 10	Direct	Leo
Aug. 26	Direct	Virgo
Sept. 12	Direct	Libra
Oct. 1	Direct	Scorpio
Oct. 25	Retrograde S	Scorpio
Nov. 15	Direct S	Scorpio
Dec. 7	Direct	Sagittarius
Dec. 27	Direct	Capricorn

1948

Jan. 14	Direct	Aquarius
Feb. 2	Direct	Pisces
Feb. 11	Retrograde S	Pisces
Feb. 20	Retrograde	Aquarius
Mar. 4	Direct S	Aquarius
Mar. 18	Direct	Pisces
Apr. 9	Direct	Aries
Apr. 25	Direct	Taurus
May 9	Direct	Gemini
May 28	Direct	Cancer
June 11	Retrograde S	Cancer
June 28	Retrograde	Gemini
July 5	Direct S	Gemini
July 12	Direct	Cancer
Aug. 2	Direct	Leo
Aug. 17	Direct	Virgo
Sept. 3	Direct	Libra
Sept. 27	Direct	Scorpio
Oct. 8	Retrograde S	Scorpio
Oct. 17	Retrograde	Libra
Oct. 29	Direct S	Libra
Nov. 10	Direct	Scorpio

| Nov. 29 | Direct | Sagittarius |
| Dec. 18 | Direct | Capricorn |

1949

Jan. 6	Direct	Aquarius
Jan. 25	Retrograde S	Aquarius
Feb. 15	Direct S	Aquarius
Mar. 14	Direct	Pisces
Apr. 1	Direct	Aries
Apr. 16	Direct	Taurus
May 2	Direct	Gemini
May 23	Retrograde S	Gemini
June 16	Direct S	Gemini
July 10	Direct	Cancer
July 25	Direct	Leo
Aug. 9	Direct	Virgo
Aug. 28	Direct	Libra
Sept. 21	Retrograde S	Libra
Oct. 13	Direct S	Libra
Nov. 3	Direct	Scorpio
Nov. 22	Direct	Sagittarius
Dec. 11	Direct	Capricorn

1950

Jan. 1	Direct	Aquarius
Jan. 8	Retrograde S	Aquarius
Jan. 15	Retrograde	Capricorn
Jan. 29	Direct S	Capricorn
Feb. 15	Direct	Aquarius
Mar. 8	Direct	Pisces
Mar. 24	Direct	Aries
Apr. 8	Direct	Taurus
May 3	Retrograde S	Taurus
May 27	Direct S	Taurus
June 14	Direct	Gemini
July 2	Direct	Cancer
July 16	Direct	Leo
Aug. 2	Direct	Virgo

Aug. 27	Direct	Libra
Sept. 4	Retrograde S	Libra
Sept. 11	Retrograde	Virgo
Sept. 26	Direct S	Virgo
Oct. 9	Direct	Libra
Oct. 27	Direct	Scorpio
Nov. 15	Direct	Sagittarius
Dec. 5	Direct	Capricorn
Dec. 23	Retrograde S	Capricorn

1951

Jan. 12	Direct S	Capricorn
Feb. 9	Direct	Aquarius
Feb. 28	Direct	Pisces
Mar. 16	Direct	Aries
Apr. 2	Direct	Taurus
Apr. 14	Retrograde S	Taurus
May 2	Retrograde	Aries
May 8	Direct S	Aries
May 15	Direct	Taurus
June 9	Direct	Gemini
June 24	Direct	Cancer
July 8	Direct	Leo
July 27	Direct	Virgo
Aug. 17	Retrograde S	Virgo
Sept. 10	Direct S	Virgo
Oct. 2	Direct	Libra
Oct. 20	Direct	Scorpio
Nov. 8	Direct	Sagittarius
Dec. 2	Direct	Capricorn
Dec. 7	Retrograde S	Capricorn
Dec. 12	Retrograde	Sagittarius
Dec. 27	Direct S	Sagittarius

1952

Jan. 13	Direct	Capricorn
Feb. 3	Direct	Aquarius
Feb. 20	Direct	Pisces

Mar. 7	Direct	Aries
Mar. 26	Retrograde S	Aries
Apr. 19	Direct S	Aries
May 14	Direct	Taurus
May 31	Direct	Gemini
June 14	Direct	Cancer
June 30	Direct	Leo
July 30	Retrograde S	Leo
Aug. 22	Direct S	Leo
Sept. 7	Direct	Virgo
Sept. 23	Direct	Libra
Oct. 11	Direct	Scorpio
Nov. 1	Direct	Sagittarius
Nov. 20	Retrograde S	Sagittarius
Dec. 10	Direct S	Sagittarius

1953

Jan. 6	Direct	Capricorn
Jan. 26	Direct	Aquarius
Feb. 12	Direct	Pisces
Mar. 3	Direct	Aries
Mar. 9	Retrograde S	Aries
Mar. 16	Retrograde	Pisces
Apr. 1	Direct S	Pisces
Apr. 17	Direct	Aries
May 8	Direct	Taurus
May 23	Direct	Gemini
June 6	Direct	Cancer
June 26	Direct	Leo
July 11	Retrograde S	Leo
July 28	Retrograde	Cancer
Aug. 5	Direct S	Cancer
Aug. 11	Direct	Leo
Aug. 31	Direct	Virgo
Sept. 16	Direct	Libra
Oct. 4	Direct	Scorpio
Oct. 31	Direct	Sagittarius
Nov. 4	Retrograde S	Sagittarius
Nov. 7	Retrograde	Scorpio

Nov. 24	Direct S	Scorpio
Dec. 10	Direct	Sagittarius
Dec. 30	Direct	Capricorn

1954

Jan. 18	Direct	Aquarius
Feb. 4	Direct	Pisces
Feb. 20	Retrograde S	Pisces
Mar. 14	Direct S	Pisces
Apr. 13	Direct	Aries
Apr. 30	Direct	Taurus
May 14	Direct	Gemini
May 30	Direct	Cancer
June 23	Retrograde S	Cancer
July 17	Direct S	Cancer
Aug. 7	Direct	Leo
Aug. 22	Direct	Virgo
Sept. 8	Direct	Libra
Sept. 29	Direct	Scorpio
Oct. 18	Retrograde S	Scorpio
Nov. 4	Retrograde	Libra
Nov. 8	Direct S	Libra
Nov. 11	Direct	Scorpio
Dec. 4	Direct	Sagittarius
Dec. 23	Direct	Capricorn

1955

Jan. 11	Direct	Aquarius
Feb. 4	Retrograde S	Aquarius
Feb. 25	Direct S	Aquarius
Mar. 18	Direct	Pisces
Apr. 6	Direct	Aries
Apr. 22	Direct	Taurus
May 6	Direct	Gemini
June 4	Retrograde S	Gemini
June 28	Direct S	Gemini
July 13	Direct	Cancer
July 30	Direct	Leo

Aug. 14	Direct	Virgo
Sept. 1	Direct	Libra
Oct. 1	Retrograde S	Libra
Oct. 23	Direct S	Libra
Nov. 8	Direct	Scorpio
Nov. 27	Direct	Sagittarius
Dec. 16	Direct	Capricorn

1956

Jan. 4	Direct	Aquarius
Jan. 18	Retrograde S	Aquarius
Feb. 2	Retrograde	Capricorn
Feb. 8	Direct S	Capricorn
Feb. 15	Direct	Aquarius
Mar. 11	Direct	Pisces
Mar. 29	Direct	Aries
Apr. 12	Direct	Taurus
Apr. 30	Direct	Gemini
May 14	Retrograde S	Gemini
June 7	Direct S	Gemini
July 7	Direct	Cancer
July 21	Direct	Leo
Aug. 6	Direct	Virgo
Aug. 26	Direct	Libra
Sept. 13	Retrograde S	Libra
Sept. 30	Retrograde	Virgo
Oct. 5	Direct S	Virgo
Oct. 11	Direct	Libra
Oct. 31	Direct	Scorpio
Nov. 19	Direct	Sagittarius
Dec. 8	Direct	Capricorn

1957

Jan. 1	Retrograde S	Capricorn
Jan. 22	Direct S	Capricorn
Feb. 12	Direct	Aquarius
Mar. 4	Direct	Pisces

Mar. 21	Direct	Aries
Apr. 5	Direct	Taurus
Apr. 25	Retrograde S	Taurus
May 19	Direct S	Taurus
June 12	Direct	Gemini
June 28	Direct	Cancer
July 13	Direct	Leo
July 30	Direct	Virgo
Aug. 27	Retrograde S	Virgo
Sept. 19	Direct S	Virgo
Oct. 6	Direct	Libra
Oct. 24	Direct	Scorpio
Nov. 11	Direct	Sagittarius
Dec. 2	Direct	Capricorn
Dec. 16	Retrograde S	Capricorn
Dec. 28	Retrograde	Sagittarius

1958

Jan. 5	Direct S	Sagittarius
Jan. 14	Direct	Capricorn
Feb. 6	Direct	Aquarius
Feb. 25	Direct	Pisces
Mar. 12	Direct	Aries
Apr. 3	Direct	Taurus
Apr. 6	Retrograde S	Taurus
Apr. 10	Retrograde	Aries
Apr. 30	Direct S	Aries
May 17	Direct	Taurus
June 6	Direct	Gemini
June 20	Direct	Cancer
July 5	Direct	Leo
July 26	Direct	Virgo
Aug. 9	Retrograde S	Virgo
Aug. 23	Retrograde	Leo
Sept. 2	Direct S	Leo
Sept. 11	Direct	Virgo
Sept. 29	Direct	Libra
Oct. 16	Direct	Scorpio

Nov. 5	Direct	Sagittarius
Nov. 30	Retrograde S	Sagittarius
Dec. 20	Direct S	Sagittarius

1959

Jan. 10	Direct	Capricorn
Jan. 30	Direct	Aquarius
Feb. 17	Direct	Pisces
Mar. 5	Direct	Aries
Mar. 19	Retrograde S	Aries
Apr. 12	Direct S	Aries
May 13	Direct	Taurus
May 28	Direct	Gemini
June 11	Direct	Cancer
June 28	Direct	Leo
July 23	Retrograde S	Leo
Aug. 16	Direct S	Leo
Sept. 5	Direct	Virgo
Sept. 21	Direct	Libra
Oct. 9	Direct	Scorpio
Oct. 31	Direct	Sagittarius
Nov. 14	Retrograde S	Sagittarius
Nov. 25	Retrograde	Scorpio
Dec. 4	Direct S	Scorpio
Dec. 13	Direct	Sagittarius

1960

Jan. 4	Direct	Capricorn
Jan. 23	Direct	Aquarius
Feb. 9	Direct	Pisces
Mar. 1	Retrograde S	Pisces
Mar. 24	Direct S	Pisces
Apr. 16	Direct	Aries
May 4	Direct	Taurus
May 19	Direct	Gemini
June 3	Direct	Cancer
July 1	Direct	Leo
July 3	Retrograde S	Leo

July 6	Retrograde	Cancer
July 27	Direct S	Cancer
Aug. 10	Direct	Leo
Aug. 27	Direct	Virgo
Sept. 12	Direct	Libra
Oct. 1	Direct	Scorpio
Oct. 27	Retrograde S	Scorpio
Nov. 17	Direct S	Scorpio
Dec. 7	Direct	Sagittarius
Dec. 27	Direct	Capricorn

1961

Jan. 14	Direct	Aquarius
Feb. 2	Direct	Pisces
Feb. 13	Retrograde S	Pisces
Feb. 25	Retrograde	Aquarius
Mar. 7	Direct S	Aquarius
Mar. 18	Direct	Pisces
Apr. 10	Direct	Aries
Apr. 26	Direct	Taurus
May 10	Direct	Gemini
May 28	Direct	Cancer
June 14	Retrograde S	Cancer
July 9	Direct S	Cancer
Aug. 4	Direct	Leo
Aug. 19	Direct	Virgo
Sept. 5	Direct	Libra
Sept. 27	Direct	Scorpio
Oct. 11	Retrograde S	Scorpio
Oct. 22	Retrograde	Libra
Oct. 31	Direct S	Libra
Nov. 11	Direct	Scorpio
Dec. 1	Direct	Sagittarius
Dec. 20	Direct	Capricorn

1962

Jan. 7	Direct	Aquarius
Jan. 27	Retrograde S	Aquarius

Feb. 18	Direct S	Aquarius
Mar. 15	Direct	Pisces
Apr. 3	Direct	Aries
Apr. 18	Direct	Taurus
May 3	Direct	Gemini
May 26	Retrograde S	Gemini
June 19	Direct S	Gemini
July 11	Direct	Cancer
July 26	Direct	Leo
Aug. 11	Direct	Virgo
Aug. 29	Direct	Libra
Sept. 24	Retrograde S	Libra
Oct. 15	Direct S	Libra
Nov. 5	Direct	Scorpio
Nov. 23	Direct	Sagittarius
Dec. 13	Direct	Capricorn

1963

Jan. 2	Direct	Aquarius
Jan. 11	Retrograde S	Aquarius
Jan. 20	Retrograde	Capricorn
Feb. 1	Direct S	Capricorn
Feb. 15	Direct	Aquarius
Mar. 9	Direct	Pisces
Mar. 26	Direct	Aries
Apr. 10	Direct	Taurus
May 3	Direct	Gemini
May 7	Retrograde S	Gemini
May 11	Retrograde	Taurus
May 30	Direct S	Taurus
June 15	Direct	Gemini
July 4	Direct	Cancer
July 18	Direct	Leo
Aug. 3	Direct	Virgo
Aug. 27	Direct	Libra
Sept. 7	Retrograde S	Libra
Sept. 17	Retrograde	Virgo
Sept. 29	Direct S	Virgo
Oct. 10	Direct	Libra

Oct. 29	Direct	Scorpio
Nov. 16	Direct	Sagittarius
Dec. 6	Direct	Capricorn
Dec. 26	Retrograde S	Capricorn

1964

Jan. 15	Direct S	Capricorn
Feb. 11	Direct	Aquarius
Mar. 1	Direct	Pisces
Mar. 17	Direct	Aries
Apr. 2	Direct	Taurus
Apr. 17	Retrograde S	Taurus
May 10	Direct S	Taurus
June 9	Direct	Gemini
June 24	Direct	Cancer
July 9	Direct	Leo
July 27	Direct	Virgo
Aug. 19	Retrograde S	Virgo
Sept. 11	Direct S	Virgo
Oct. 3	Direct	Libra
Oct. 20	Direct	Scorpio
Nov. 8	Direct	Sagittarius
Dec. 1	Direct	Capricorn
Dec. 9	Retrograde S	Capricorn
Dec. 16	Retrograde	Sagittarius
Dec. 29	Direct S	Sagittarius

1965

Jan. 13	Direct	Capricorn
Feb. 3	Direct	Aquarius
Feb. 21	Direct	Pisces
Mar. 9	Direct	Aries
Mar. 29	Retrograde S	Aries
Apr. 22	Direct S	Aries
May 15	Direct	Taurus
June 2	Direct	Gemini
June 16	Direct	Cancer
July 1	Direct	Leo

July 31	Direct	Virgo
Aug. 2	Retrograde S	Virgo
Aug. 3	Retrograde	Leo
Aug. 25	Direct S	Leo
Sept. 8	Direct	Virgo
Sept. 25	Direct	Libra
Oct. 13	Direct	Scorpio
Nov. 2	Direct	Sagittarius
Nov. 23	Retrograde S	Sagittarius
Dec. 13	Direct S	Sagittarius

1966

Jan. 7	Direct	Capricorn
Jan. 27	Direct	Aquarius
Feb. 13	Direct	Pisces
Mar. 3	Direct	Aries
Mar. 12	Retrograde S	Aries
Mar. 22	Retrograde	Pisces
Apr. 4	Direct S	Pisces
Apr. 18	Direct	Aries
May 9	Direct	Taurus
May 24	Direct	Gemini
June 8	Direct	Cancer
June 27	Direct	Leo
July 15	Retrograde S	Leo
Aug. 8	Direct S	Leo
Sept. 1	Direct	Virgo
Sept. 17	Direct	Libra
Oct. 6	Direct	Scorpio
Oct. 30	Direct	Sagittarius
Nov. 6	Retrograde S	Sagittarius
Nov. 13	Retrograde	Scorpio
Nov. 26	Direct S	Scorpio
Dec. 11	Direct	Sagittarius

1967

Jan. 1	Direct	Capricorn
Jan. 19	Direct	Aquarius

Feb. 6	Direct	Pisces
Feb. 23	Retrograde S	Pisces
Mar. 17	Direct S	Pisces
Apr. 14	Direct	Aries
May 2	Direct	Taurus
May 16	Direct	Gemini
May 31	Direct	Cancer
June 26	Retrograde S	Cancer
July 20	Direct S	Cancer
Aug. 9	Direct	Leo
Aug. 24	Direct	Virgo
Sept. 9	Direct	Libra
Sept. 30	Direct	Scorpio
Oct. 21	Retrograde S	Scorpio
Nov. 10	Direct S	Scorpio
Dec. 5	Direct	Sagittarius
Dec. 25	Direct	Capricorn

1968

Jan. 12	Direct	Aquarius
Feb. 1	Direct	Pisces
Feb. 6	Retrograde S	Pisces
Feb. 11	Retrograde	Aquarius
Feb. 28	Direct S	Aquarius
Mar. 17	Direct	Pisces
Apr. 7	Direct	Aries
Apr. 22	Direct	Taurus
May 7	Direct	Gemini
May 30	Direct	Cancer
June 6	Retrograde S	Cancer
June 14	Retrograde	Gemini
June 30	Direct S	Gemini
July 13	Direct	Cancer
July 31	Direct	Leo
Aug. 15	Direct	Virgo
Sept. 1	Direct	Libra
Sept. 28	Direct	Scorpio
Oct. 3	Retrograde S	Scorpio
Oct. 8	Retrograde	Libra

Oct. 24	Direct S	Libra
Nov. 8	Direct	Scorpio
Nov. 27	Direct	Sagittarius
Dec. 16	Direct	Capricorn

1969

Jan. 4	Direct	Aquarius
Jan. 20	Retrograde S	Aquarius
Feb. 10	Direct S	Aquarius
Mar. 12	Direct	Pisces
Mar. 30	Direct	Aries
Apr. 14	Direct	Taurus
Apr. 30	Direct	Gemini
May 18	Retrograde S	Gemini
June 10	Direct S	Gemini
July 8	Direct	Cancer
July 23	Direct	Leo
Aug. 7	Direct	Virgo
Aug. 27	Direct	Libra
Sept. 16	Retrograde S	Libra
Oct. 7	Retrograde	Virgo
Oct. 8	Direct S	Virgo
Oct. 9	Direct	Libra
Nov. 1	Direct	Scorpio
Nov. 20	Direct	Sagittarius
Dec. 9	Direct	Capricorn

1970

Jan. 4	Retrograde S	Aquarius
Jan. 24	Direct S	Capricorn
Feb. 13	Direct	Aquarius
Mar. 6	Direct	Pisces
Mar. 22	Direct	Aries
Apr. 6	Direct	Taurus
Apr. 28	Retrograde S	Taurus
May 22	Direct S	Taurus
June 13	Direct	Gemini
June 30	Direct	Cancer

July 14	Direct	Leo
July 31	Direct	Virgo
Aug. 30	Retrograde S	Virgo
Sept. 22	Direct S	Virgo
Oct. 7	Direct	Libra
Oct. 25	Direct	Scorpio
Nov. 13	Direct	Sagittarius
Dec. 3	Direct	Capricorn
Dec. 19	Retrograde S	Capricorn

1971

Jan. 3	Retrograde	Sagittarius
Jan. 8	Direct S	Sagittarius
Jan. 14	Direct	Capricorn
Feb. 8	Direct	Aquarius
Feb. 26	Direct	Pisces
Mar. 14	Direct	Aries
Apr. 1	Direct	Taurus
Apr. 9	Retrograde S	Taurus
Apr. 19	Retrograde	Aries
May 3	Direct S	Aries
May 17	Direct	Taurus
June 7	Direct	Gemini
June 21	Direct	Cancer
July 6	Direct	Leo
July 26	Direct	Virgo
Aug. 13	Retrograde S	Virgo
Aug. 30	Retrograde	Leo
Sept. 5	Direct S	Leo
Sept. 11	Direct	Virgo
Sept. 30	Direct	Libra
Oct. 17	Direct	Scorpio
Nov. 6	Direct	Sagittarius
Dec. 3	Retrograde S	Sagittarius
Dec. 23	Direct S	Sagittarius

1972

Jan. 11	Direct	Capricorn
Feb. 1	Direct	Aquarius
Feb. 18	Direct	Pisces
Mar. 5	Direct	Aries
Mar. 21	Retrograde S	Aries
Apr. 14	Direct S	Aries
May 13	Direct	Taurus
May 29	Direct	Gemini
June 12	Direct	Cancer
June 28	Direct	Leo
July 25	Retrograde S	Leo
Aug. 18	Direct S	Leo
Sept. 5	Direct	Virgo
Sept. 21	Direct	Libra
Oct. 9	Direct	Scorpio
Oct. 31	Direct	Sagittarius
Nov. 16	Retrograde S	Sagittarius
Nov. 29	Retrograde	Scorpio
Dec. 5	Direct S	Scorpio
Dec. 13	Direct	Sagittarius

1973

Jan. 4	Direct	Capricorn
Jan. 23	Direct	Aquarius
Feb. 10	Direct	Pisces
Mar. 4	Retrograde S	Pisces
Mar. 27	Direct S	Pisces
Apr. 17	Direct	Aries
May 6	Direct	Taurus
May 20	Direct	Gemini
June 4	Direct	Cancer
June 27	Direct	Leo
July 6	Retrograde S	Leo
July 16	Retrograde	Cancer
July 31	Direct S	Cancer
Aug. 11	Direct	Leo
Aug. 28	Direct	Virgo

Sept. 13	Direct	Libra
Oct. 3	Direct	Scorpio
Oct. 30	Retrograde S	Scorpio
Nov. 19	Direct S	Scorpio
Dec. 9	Direct	Sagittarius
Dec. 28	Direct	Capricorn

1974

Jan. 16	Direct	Aquarius
Feb. 3	Direct	Pisces
Feb. 16	Retrograde S	Pisces
Mar. 2	Retrograde	Aquarius
Mar. 10	Direct S	Aquarius
Mar. 18	Direct	Pisces
Apr. 11	Direct	Aries
Apr. 28	Direct	Taurus
May 12	Direct	Gemini
May 29	Direct	Cancer
June 18	Retrograde S	Cancer
July 12	Direct S	Cancer
Aug. 5	Direct	Leo
Aug. 20	Direct	Virgo
Sept. 6	Direct	Libra
Sept. 28	Direct	Scorpio
Oct. 14	Retrograde S	Scorpio
Oct. 27	Retrograde	Libra
Nov. 3	Direct S	Libra
Nov. 11	Direct	Scorpio
Dec. 2	Direct	Sagittarius
Dec. 21	Direct	Capricorn

1975

Jan. 9	Direct	Aquarius
Jan. 30	Retrograde S	Aquarius
Feb. 21	Direct S	Aquarius
Mar. 16	Direct	Pisces
Apr. 4	Direct	Aries
Apr. 19	Direct	Taurus

May 4	Direct	Gemini
May 29	Retrograde S	Gemini
June 22	Direct S	Gemini
July 12	Direct	Cancer
July 28	Direct	Leo
Aug. 12	Direct	Virgo
Aug. 30	Direct	Libra
Sept. 27	Retrograde S	Libra
Oct. 18	Direct S	Libra
Nov. 6	Direct	Scorpio
Nov. 25	Direct	Sagittarius
Dec. 14	Direct	Capricorn

1976

Jan. 3	Direct	Aquarius
Jan. 14	Retrograde S	Aquarius
Jan. 25	Retrograde	Capricorn
Feb. 4	Direct S	Capricorn
Feb. 16	Direct	Aquarius
Mar. 9	Direct	Pisces
Mar. 26	Direct	Aries
Apr. 10	Direct	Taurus
Apr. 30	Direct	Gemini
May 9	Retrograde S	Gemini
May 20	Retrograde	Taurus
June 2	Direct S	Taurus
June 14	Direct	Gemini
July 4	Direct	Cancer
July 19	Direct	Leo
Aug. 3	Direct	Virgo
Aug. 26	Direct	Libra
Sept. 9	Retrograde S	Libra
Sept. 21	Retrograde	Virgo
Oct. 1	Direct S	Virgo
Oct. 10	Direct	Libra
Oct. 29	Direct	Scorpio
Nov. 17	Direct	Sagittarius
Dec. 6	Direct	Capricorn
Dec. 28	Retrograde S	Capricorn

1977

Jan. 17	Direct S	Capricorn
Feb. 11	Direct	Aquarius
Mar. 2	Direct	Pisces
Mar. 18	Direct	Aries
Apr. 3	Direct	Taurus
Apr. 20	Retrograde S	Taurus
May 14	Direct S	Taurus
June 11	Direct	Gemini
June 26	Direct	Cancer
July 10	Direct	Leo
July 28	Direct	Virgo
Aug. 22	Retrograde S	Virgo
Sept. 14	Direct S	Virgo
Oct. 4	Direct	Libra
Oct. 21	Direct	Scorpio
Nov. 9	Direct	Sagittarius
Dec. 1	Direct	Capricorn
Dec. 12	Retrograde S	Capricorn
Dec. 21	Retrograde	Sagittarius

1978

Jan. 1	Direct S	Sagittarius
Jan. 14	Direct	Capricorn
Feb. 4	Direct	Aquarius
Feb. 22	Direct	Pisces
Mar. 10	Direct	Aries
Apr. 1	Retrograde S	Aries
Apr. 25	Direct S	Aries
May 16	Direct	Taurus
June 3	Direct	Gemini
June 17	Direct	Cancer
July 3	Direct	Leo
July 27	Direct	Virgo
Aug. 5	Retrograde S	Virgo
Aug. 13	Retrograde	Leo
Aug. 28	Direct S	Leo
Sept. 10	Direct	Virgo

Sept. 26	Direct	Libra
Oct. 14	Direct	Scorpio
Nov. 3	Direct	Sagittarius
Nov. 26	Retrograde S	Sagittarius
Dec. 15	Direct S	Sagittarius

1979

Jan. 9	Direct	Capricorn
Jan. 28	Direct	Aquarius
Feb. 15	Direct	Pisces
Mar. 4	Direct	Aries
Mar. 15	Retrograde S	Aries
Mar. 28	Retrograde	Pisces
Apr. 7	Direct S	Pisces
Apr. 17	Direct	Aries
May 11	Direct	Taurus
May 26	Direct	Gemini
June 9	Direct	Cancer
June 27	Direct	Leo
July 18	Retrograde S	Leo
Aug. 11	Direct S	Leo
Sept. 3	Direct	Virgo
Sept. 18	Direct	Libra
Oct. 7	Direct	Scorpio
Oct. 30	Direct	Sagittarius
Nov. 9	Retrograde S	Sagittarius
Nov. 18	Retrograde	Scorpio
Nov. 29	Direct S	Scorpio
Dec. 12	Direct	Sagittarius

1980

Jan. 2	Direct	Capricorn
Jan. 21	Direct	Aquarius
Feb. 7	Direct	Pisces
Feb. 26	Retrograde S	Pisces
Mar. 19	Direct S	Pisces
Apr. 14	Direct	Aries
May 2	Direct	Taurus

May 16	Direct	Gemini
June 1	Direct	Cancer
June 28	Retrograde S	Cancer
July 22	Direct S	Cancer
Aug. 9	Direct	Leo
Aug. 24	Direct	Virgo
Sept. 10	Direct	Libra
Sept. 30	Direct	Scorpio
Oct. 23	Retrograde S	Scorpio
Nov. 12	Direct S	Scorpio
Dec. 6	Direct	Sagittarius
Dec. 25	Direct	Capricorn

1981

Jan. 12	Direct	Aquarius
Jan. 31	Direct	Pisces
Feb. 8	Retrograde S	Pisces
Feb. 16	Retrograde	Aquarius
Mar. 2	Direct S	Aquarius
Mar. 18	Direct	Pisces
Apr. 8	Direct	Aries
Apr. 24	Direct	Taurus
May 8	Direct	Gemini
May 28	Direct	Cancer
June 9	Retrograde S	Cancer
June 23	Retrograde	Gemini
July 3	Direct S	Gemini
July 13	Direct	Cancer
Aug. 1	Direct	Leo
Aug. 16	Direct	Virgo
Sept. 3	Direct	Libra
Sept. 27	Direct	Scorpio
Oct. 6	Retrograde S	Scorpio
Oct. 14	Retrograde	Libra
Oct. 27	Direct S	Libra
Nov. 9	Direct	Scorpio
Nov. 29	Direct	Sagittarius
Dec. 18	Direct	Capricorn

1982

Jan. 5	Direct	Aquarius
Jan. 23	Retrograde S	Aquarius
Feb. 13	Direct S	Aquarius
Mar. 14	Direct	Pisces
Apr. 1	Direct	Aries
Apr. 15	Direct	Taurus
May 1	Direct	Gemini
May 21	Retrograde S	Gemini
June 14	Direct S	Gemini
July 8	Direct	Cancer
July 24	Direct	Leo
Aug. 8	Direct	Virgo
Aug. 28	Direct	Libra
Sept. 19	Retrograde S	Libra
Oct. 11	Direct S	Libra
Nov. 3	Direct	Scorpio
Nov. 21	Direct	Sagittarius
Dec. 11	Direct	Capricorn

1983

Jan. 1	Direct	Aquarius
Jan. 7	Retrograde S	Aquarius
Jan. 12	Retrograde	Capricorn
Jan. 27	Direct S	Capricorn
Feb. 14	Direct	Aquarius
Mar. 7	Direct	Pisces
Mar. 24	Direct	Aries
Apr. 7	Direct	Taurus
May 1	Retrograde S	Taurus
May 25	Direct S	Taurus
June 14	Direct	Gemini
July 2	Direct	Cancer
July 16	Direct	Leo
Aug. 1	Direct	Virgo
Aug. 29	Direct	Libra
Sept. 2	Retrograde S	Libra
Sept. 6	Retrograde	Virgo

Sept. 25	Direct S	Virgo
Oct. 9	Direct	Libra
Oct. 26	Direct	Scorpio
Nov. 14	Direct	Sagittarius
Dec. 4	Direct	Capricorn
Dec. 22	Retrograde S	Capricorn

1984

Jan. 11	Direct S	Capricorn
Feb. 9	Direct	Aquarius
Feb. 27	Direct	Pisces
Mar. 14	Direct	Aries
Apr. 1	Direct	Taurus
Apr. 12	Retrograde S	Taurus
Apr. 25	Retrograde	Aries
May 5	Direct S	Aries
May 15	Direct	Taurus
June 7	Direct	Gemini
June 22	Direct	Cancer
July 6	Direct	Leo
July 26	Direct	Virgo
Aug. 15	Retrograde S	Virgo
Sept. 7	Direct S	Virgo
Oct. 1	Direct	Libra
Oct. 18	Direct	Scorpio
Nov. 6	Direct	Sagittarius
Dec. 1	Direct	Capricorn
Dec. 5	Retrograde S	Capricorn
Dec. 8	Retrograde	Sagittarius
Dec. 24	Direct S	Sagittarius

1985

Jan. 11	Direct	Capricorn
Feb. 1	Direct	Aquarius
Feb. 19	Direct	Pisces
Mar. 7	Direct	Aries
Mar. 25	Retrograde S	Aries
Apr. 17	Direct S	Aries

May 14	Direct	Taurus
May 31	Direct	Gemini
June 13	Direct	Cancer
June 30	Direct	Leo
July 28	Retrograde S	Leo
Aug. 21	Direct S	Leo
Sept. 7	Direct	Virgo
Sept. 23	Direct	Libra
Oct. 10	Direct	Scorpio
Oct. 31	Direct	Sagittarius
Nov. 18	Retrograde S	Sagittarius
Dec. 5	Retrograde	Scorpio
Dec. 8	Direct S	Scorpio
Dec. 12	Direct	Sagittarius

1986

Jan. 6	Direct	Capricorn
Jan. 25	Direct	Aquarius
Feb. 11	Direct	Pisces
Mar. 3	Direct	Aries
Mar. 7	Retrograde S	Aries
Mar. 11	Retrograde	Pisces
Mar. 30	Direct S	Pisces
Apr. 17	Direct	Aries
May 7	Direct	Taurus
May 22	Direct	Gemini
June 5	Direct	Cancer
June 26	Direct	Leo
July 10	Retrograde S	Leo
July 24	Retrograde	Cancer
Aug. 3	Direct S	Cancer
Aug. 12	Direct	Leo
Aug. 30	Direct	Virgo
Sept. 15	Direct	Libra
Oct. 4	Direct	Scorpio
Nov. 2	Retrograde S	Scorpio
Nov. 22	Direct S	Scorpio
Dec. 10	Direct	Sagittarius
Dec. 30	Direct	Capricorn

1987

Jan. 17	Direct	Aquarius
Feb. 4	Direct	Pisces
Feb. 18	Retrograde S	Pisces
Mar. 12	Retrograde	Aquarius
Mar. 13	Direct S	Aquarius
Mar. 14	Direct	Pisces
Apr. 13	Direct	Aries
Apr. 29	Direct	Taurus
May 13	Direct	Gemini
May 30	Direct	Cancer
June 21	Retrograde S	Cancer
July 15	Direct S	Cancer
Aug. 7	Direct	Leo
Aug. 22	Direct	Virgo
Sept. 7	Direct	Libra
Sept. 28	Direct	Scorpio
Oct. 16	Retrograde S	Scorpio
Nov. 1	Retrograde	Libra
Nov. 6	Direct S	Libra
Nov. 12	Direct	Scorpio
Dec. 3	Direct	Sagittarius
Dec. 22	Direct	Capricorn

1988

Jan. 10	Direct	Aquarius
Feb. 2	Retrograde S	Aquarius
Feb. 23	Direct S	Aquarius
Mar. 16	Direct	Pisces
Apr. 5	Direct	Aries
Apr. 20	Direct	Taurus
May 5	Direct	Gemini
June 1	Retrograde S	Gemini
June 25	Direct S	Gemini
July 12	Direct	Cancer
July 29	Direct	Leo
Aug. 12	Direct	Virgo
Aug. 31	Direct	Libra

Sept. 29	Retrograde S	Libra
Oct. 20	Direct S	Libra
Nov. 6	Direct	Scorpio
Nov. 25	Direct	Sagittarius
Dec. 14	Direct	Capricorn

1989

Jan. 3	Direct	Aquarius
Jan. 16	Retrograde S	Aquarius
Jan. 29	Retrograde	Capricorn
Feb. 6	Direct S	Capricorn
Feb. 14	Direct	Aquarius
Mar. 10	Direct	Pisces
Mar. 28	Direct	Aries
Apr. 12	Direct	Taurus
Apr. 30	Direct	Gemini
May 12	Retrograde S	Gemini
May 29	Retrograde	Taurus
June 5	Direct S	Taurus
June 12	Direct	Gemini
July 6	Direct	Cancer
July 20	Direct	Leo
Aug. 5	Direct	Virgo
Aug. 26	Direct	Libra
Sept. 12	Retrograde S	Libra
Sept. 26	Retrograde	Virgo
Oct. 4	Direct S	Virgo
Oct. 11	Direct	Libra
Oct. 30	Direct	Scorpio
Nov. 18	Direct	Sagittarius
Dec. 7	Direct	Capricorn
Dec. 31	Retrograde S	Capricorn

1990

Jan. 20	Direct S	Capricorn
Feb. 12	Direct	Aquarius
Mar. 3	Direct	Pisces
Mar. 20	Direct	Aries

Apr. 4	Direct	Taurus
Apr. 23	Retrograde S	Taurus
May 17	Direct S	Taurus
June 12	Direct	Gemini
June 28	Direct	Cancer
July 12	Direct	Leo
July 29	Direct	Virgo
Aug. 25	Retrograde S	Virgo
Sept. 17	Direct S	Virgo
Oct. 5	Direct	Libra
Oct. 23	Direct	Scorpio
Nov. 11	Direct	Sagittarius
Dec. 2	Direct	Capricorn
Dec. 15	Retrograde S	Capricorn
Dec. 26	Retrograde	Sagittarius

1991

Jan. 3	Direct S	Sagittarius
Jan. 14	Direct	Capricorn
Feb. 6	Direct	Aquarius
Feb. 24	Direct	Pisces
Mar. 12	Direct	Aries
Apr. 4	Retrograde S	Aries
Apr. 28	Direct S	Aries
May 17	Direct	Taurus
June 5	Direct	Gemini
June 19	Direct	Cancer
July 4	Direct	Leo
July 26	Direct	Virgo
Aug. 8	Retrograde S	Virgo
Aug. 20	Retrograde	Leo
Aug. 31	Direct S	Leo
Sept. 10	Direct	Virgo
Sept. 28	Direct	Libra
Oct. 15	Direct	Scorpio
Nov. 4	Direct	Sagittarius
Nov. 28	Retrograde S	Sagittarius
Dec. 18	Direct S	Sagittarius

1992

Jan. 10	Direct	Capricorn
Jan. 30	Direct	Aquarius
Feb. 16	Direct	Pisces
Mar. 4	Direct	Aries
Mar. 17	Retrograde S	Aries
Apr. 4	Retrograde	Pisces
Apr. 9	Direct S	Pisces
Apr. 14	Direct	Aries
May 11	Direct	Taurus
May 27	Direct	Gemini
June 9	Direct	Cancer
June 27	Direct	Leo
July 20	Retrograde S	Leo
Aug. 13	Direct S	Leo
Sept. 3	Direct	Virgo
Sept. 19	Direct	Libra
Oct. 7	Direct	Scorpio
Oct. 29	Direct	Sagittarius
Nov. 11	Retrograde S	Sagittarius
Nov. 22	Retrograde	Scorpio
Dec. 1	Direct S	Scorpio
Dec. 12	Direct	Sagittarius

1993

Jan. 2	Direct	Capricorn
Jan. 21	Direct	Aquarius
Feb. 7	Direct	Pisces
Feb. 28	Retrograde S	Pisces
Mar. 22	Direct S	Pisces
Apr. 15	Direct	Aries
May 4	Direct	Taurus
May 18	Direct	Gemini
June 2	Direct	Cancer
July 1	Retrograde S	Cancer
July 26	Direct S	Cancer
Aug. 10	Direct	Leo
Aug. 26	Direct	Virgo

Sept. 11	Direct	Libra
Oct. 1	Direct	Scorpio
Oct. 26	Retrograde S	Scorpio
Nov. 15	Direct S	Scorpio
Dec. 7	Direct	Sagittarius
Dec. 26	Direct	Capricorn

1994

Jan. 14	Direct	Aquarius
Feb. 1	Direct	Pisces
Feb. 11	Retrograde S	Pisces
Feb. 21	Retrograde	Aquarius
Mar. 5	Direct S	Aquarius
Mar. 18	Direct	Pisces
Apr. 9	Direct	Aries
Apr. 25	Direct	Taurus
May 10	Direct	Gemini
May 28	Direct	Cancer
June 12	Retrograde S	Cancer
July 3	Retrograde	Gemini
July 7	Direct S	Gemini
July 10	Direct	Cancer
Aug. 3	Direct	Leo
Aug. 18	Direct	Virgo
Sept. 4	Direct	Libra
Sept. 27	Direct	Scorpio
Oct. 9	Retrograde S	Scorpio
Oct. 19	Retrograde	Libra
Oct. 30	Direct S	Libra
Nov. 10	Direct	Scorpio
Nov. 30	Direct	Sagittarius
Dec. 19	Direct	Capricorn

1995

Jan. 7	Direct	Aquarius
Jan. 26	Retrograde S	Aquarius
Feb. 16	Direct S	Aquarius
Mar. 15	Direct	Pisces

Apr. 2	Direct	Aries
Apr. 17	Direct	Taurus
May 2	Direct	Gemini
May 24	Retrograde S	Gemini
June 17	Direct S	Gemini
July 10	Direct	Cancer
July 26	Direct	Leo
Aug. 10	Direct	Virgo
Aug. 29	Direct	Libra
Sept. 22	Retrograde S	Libra
Oct. 14	Direct S	Libra
Nov. 4	Direct	Scorpio
Nov. 23	Direct	Sagittarius
Dec. 12	Direct	Capricorn

1996

Jan. 1	Direct	Aquarius
Jan. 10	Retrograde S	Aquarius
Jan. 17	Retrograde	Capricorn
Jan. 30	Direct S	Capricorn
Feb. 15	Direct	Aquarius
Mar. 7	Direct	Pisces
Mar. 24	Direct	Aries
Apr. 8	Direct	Taurus
May 4	Retrograde S	Taurus
May 28	Direct S	Taurus
June 14	Direct	Gemini
July 2	Direct	Cancer
July 16	Direct	Leo
Aug. 1	Direct	Virgo
Aug. 26	Direct	Libra
Sept. 4	Retrograde S	Libra
Sept. 12	Retrograde	Virgo
Sept. 26	Direct S	Virgo
Oct. 9	Direct	Libra
Oct. 27	Direct	Scorpio
Nov. 14	Direct	Sagittarius
Dec. 4	Direct	Capricorn
Dec. 24	Retrograde S	Capricorn

1997

Jan. 13	Direct S	Capricorn
Feb. 9	Direct	Aquarius
Feb. 28	Direct	Pisces
Mar. 16	Direct	Aries
Apr. 1	Direct	Taurus
Apr. 15	Retrograde S	Taurus
May 5	Retrograde	Aries
May 8	Direct S	Aries
May 12	Direct	Taurus
June 9	Direct	Gemini
June 24	Direct	Cancer
July 8	Direct	Leo
July 27	Direct	Virgo
Aug. 18	Retrograde S	Virgo
Sept. 10	Direct S	Virgo
Oct. 2	Direct	Libra
Oct. 19	Direct	Scorpio
Nov. 7	Direct	Sagittarius
Dec. 1	Direct	Capricorn
Dec. 7	Retrograde S	Capricorn
Dec. 13	Retrograde	Sagittarius
Dec. 27	Direct S	Sagittarius

1998

Jan. 12	Direct	Capricorn
Feb. 2	Direct	Aquarius
Feb. 20	Direct	Pisces
Mar. 8	Direct	Aries
Mar. 28	Retrograde S	Aries
Apr. 20	Direct S	Aries
May 15	Direct	Taurus
June 1	Direct	Gemini
June 15	Direct	Cancer
July 1	Direct	Leo
July 31	Retrograde S	Leo
Aug. 24	Direct S	Leo
Sept. 8	Direct	Virgo

Sept. 24	Direct	Libra
Oct. 12	Direct	Scorpio
Nov. 1	Direct	Sagittarius
Nov. 21	Retrograde S	Sagittarius
Dec. 11	Direct S	Sagittarius

1999

Jan. 7	Direct	Capricorn
Jan. 26	Direct	Aquarius
Feb. 12	Direct	Pisces
Mar. 3	Direct	Aries
Mar. 10	Retrograde S	Aries
Mar. 18	Retrograde	Pisces
Apr. 2	Direct S	Pisces
Apr. 18	Direct	Aries
May 9	Direct	Taurus
May 24	Direct	Gemini
June 7	Direct	Cancer
June 26	Direct	Leo
July 13	Retrograde S	Leo
July 31	Retrograde	Cancer
Aug. 6	Direct S	Cancer
Aug. 11	Direct	Leo
Aug. 31	Direct	Virgo
Sept. 16	Direct	Libra
Oct. 5	Direct	Scorpio
Oct. 31	Direct	Sagittarius
Nov. 5	Retrograde S	Sagittarius
Nov. 10	Retrograde	Scorpio
Nov. 25	Direct S	Scorpio
Dec. 11	Direct	Sagittarius
Dec. 31	Direct	Capricorn

2000

Jan. 19	Direct	Aquarius
Feb. 5	Direct	Pisces
Feb. 21	Retrograde S	Pisces
Mar. 15	Direct S	Pisces

Apr. 13	Direct	Aries
Apr. 30	Direct	Taurus
May 14	Direct	Gemini
May 30	Direct	Cancer
June 23	Retrograde S	Cancer
July 17	Direct S	Cancer
Aug. 7	Direct	Leo
Aug. 22	Direct	Virgo
Sept. 8	Direct	Libra
Sept. 28	Direct	Scorpio
Oct. 18	Retrograde S	Scorpio
Nov. 7	Retrograde	Libra
Nov. 8	Direct S	Libra
Nov. 9	Direct	Scorpio
Dec. 4	Direct	Sagittarius
Dec. 23	Direct	Capricorn

2001

Jan. 10	Direct	Aquarius
Feb. 1	Direct	Pisces
Feb. 4	Retrograde S	Pisces
Feb. 7	Retrograde	Aquarius
Feb. 25	Direct S	Aquarius
Mar. 17	Direct	Pisces
Apr. 6	Direct	Aries
Apr. 22	Direct	Taurus
May 6	Direct	Gemini
June 4	Retrograde S	Gemini
June 28	Direct S	Gemini
July 13	Direct	Cancer
July 30	Direct	Leo
Aug. 14	Direct	Virgo
Sept. 1	Direct	Libra
Oct. 2	Retrograde S	Libra
Oct. 23	Direct S	Libra
Nov. 8	Direct	Scorpio
Nov. 26	Direct	Sagittarius
Dec. 16	Direct	Capricorn

2002

Jan. 4	Direct	Aquarius
Jan. 19	Retrograde S	Aquarius
Feb. 4	Retrograde	Capricorn
Feb. 8	Direct S	Capricorn
Feb. 13	Direct	Aquarius
Mar. 12	Direct	Pisces
Mar. 29	Direct	Aries
Apr. 13	Direct	Taurus
Apr. 30	Direct	Gemini
May 15	Retrograde S	Gemini
June 8	Direct S	Gemini
July 7	Direct	Cancer
July 22	Direct	Leo
Aug. 6	Direct	Virgo
Aug. 27	Direct	Libra
Sept. 15	Retrograde S	Libra
Oct. 2	Retrograde	Virgo
Oct. 7	Direct S	Virgo
Oct. 11	Direct	Libra
Nov. 1	Direct	Scorpio
Nov. 19	Direct	Sagittarius
Dec. 9	Direct	Capricorn

2003

Jan. 2	Retrograde S	Capricorn
Jan. 23	Direct S	Capricorn
Feb. 13	Direct	Aquarius
Mar. 5	Direct	Pisces
Mar. 21	Direct	Aries
Apr. 5	Direct	Taurus
Apr. 26	Retrograde S	Taurus
May 20	Direct S	Taurus
June 13	Direct	Gemini
June 29	Direct	Cancer
July 13	Direct	Leo
July 30	Direct	Virgo
Aug. 28	Retrograde S	Virgo

Sept. 20	Direct S	Virgo
Oct. 7	Direct	Libra
Oct. 24	Direct	Scorpio
Nov. 12	Direct	Sagittarius
Dec. 3	Direct	Capricorn
Dec. 17	Retrograde S	Capricorn
Dec. 31	Retrograde	Sagittarius

2004

Jan. 6	Direct S	Sagittarius
Jan. 14	Direct	Capricorn
Feb. 7	Direct	Aquarius
Feb. 25	Direct	Pisces
Mar. 12	Direct	Aries
Apr. 1	Direct	Taurus
Apr. 7	Retrograde S	Taurus
Apr. 13	Retrograde	Aries
Apr. 30	Direct S	Aries
May 16	Direct	Taurus
June 5	Direct	Gemini
June 20	Direct	Cancer
July 4	Direct	Leo
July 25	Direct	Virgo
Aug. 10	Retrograde S	Virgo
Aug. 25	Retrograde	Leo
Sept. 2	Direct S	Leo
Sept. 10	Direct	Virgo
Sept. 28	Direct	Libra
Oct. 16	Direct	Scorpio
Nov. 4	Direct	Sagittarius
Nov. 30	Retrograde S	Sagittarius
Dec. 20	Direct S	Sagittarius

2005

Jan. 10	Direct	Capricorn
Jan. 30	Direct	Aquarius
Feb. 16	Direct	Pisces
Mar. 5	Direct	Aries

Mar. 20	Retrograde S	Aries
Apr. 12	Direct S	Aries
May 12	Direct	Taurus
May 28	Direct	Gemini
June 11	Direct	Cancer
June 28	Direct	Leo
July 23	Retrograde S	Leo
Aug. 16	Direct S	Leo
Sept. 4	Direct	Virgo
Sept. 20	Direct	Libra
Oct. 8	Direct	Scorpio
Oct. 30	Direct	Sagittarius
Nov. 14	Retrograde S	Sagittarius
Nov. 26	Retrograde	Scorpio
Dec. 4	Direct S	Scorpio
Dec. 13	Direct	Sagittarius

2006

Jan. 4	Direct	Capricorn
Jan. 23	Direct	Aquarius
Feb. 9	Direct	Pisces
Mar. 3	Retrograde S	Pisces
Mar. 25	Direct S	Pisces
Apr. 16	Direct	Aries
May 5	Direct	Taurus
May 20	Direct	Gemini
June 3	Direct	Cancer
June 29	Direct	Leo
July 5	Retrograde S	Leo
July 11	Retrograde	Cancer
July 29	Direct S	Cancer
Aug. 11	Direct	Leo
Aug. 28	Direct	Virgo
Sept. 13	Direct	Libra
Oct. 2	Direct	Scorpio
Oct. 29	Retrograde S	Scorpio
Nov. 18	Direct S	Scorpio
Dec. 8	Direct	Sagittarius
Dec. 28	Direct	Capricorn

2007

Jan. 15	Direct	Aquarius
Feb. 2	Direct	Pisces
Feb. 14	Retrograde S	Pisces
Feb. 27	Retrograde	Aquarius
Mar. 8	Direct S	Aquarius
Mar. 18	Direct	Pisces
Apr. 11	Direct	Aries
Apr. 27	Direct	Taurus
May 11	Direct	Gemini
May 29	Direct	Cancer
June 16	Retrograde S	Cancer
July 10	Direct S	Cancer
Aug. 4	Direct	Leo
Aug. 19	Direct	Virgo
Sept. 5	Direct	Libra
Sept. 27	Direct	Scorpio
Oct. 12	Retrograde S	Scorpio
Oct. 24	Retrograde	Libra
Nov. 2	Direct S	Libra
Nov. 11	Direct	Scorpio
Dec. 1	Direct	Sagittarius
Dec. 20	Direct	Capricorn

2008

Jan. 8	Direct	Aquarius
Jan. 29	Retrograde S	Aquarius
Feb. 19	Direct S	Aquarius
Mar. 15	Direct	Pisces
Apr. 2	Direct	Aries
Apr. 18	Direct	Taurus
May 3	Direct	Gemini
May 26	Retrograde S	Gemini
June 19	Direct S	Gemini
July 11	Direct	Cancer
July 26	Direct	Leo
Aug. 10	Direct	Virgo
Aug. 29	Direct	Libra

Sept. 24	Retrograde S	Libra
Oct. 16	Direct S	Libra
Nov. 4	Direct	Scorpio
Nov. 23	Direct	Sagittarius
Dec. 12	Direct	Capricorn

2009

Jan. 1	Direct	Aquarius
Jan. 11	Retrograde S	Aquarius
Jan. 21	Retrograde	Capricorn
Feb. 1	Direct S	Capricorn
Feb. 14	Direct	Aquarius
Mar. 8	Direct	Pisces
Mar. 26	Direct	Aries
Apr. 9	Direct	Taurus
May 1	Direct	Gemini
May 7	Retrograde S	Gemini
May 14	Retrograde	Taurus
May 31	Direct S	Taurus
June 14	Direct	Gemini
July 4	Direct	Cancer
July 18	Direct	Leo
Aug. 3	Direct	Virgo
Aug. 26	Direct	Libra
Sept. 7	Retrograde S	Libra
Sept. 18	Retrograde	Virgo
Sept. 29	Direct S	Virgo
Oct. 10	Direct	Libra
Oct. 28	Direct	Scorpio
Nov. 16	Direct	Sagittarius
Dec. 5	Direct	Capricorn
Dec. 26	Retrograde S	Capricorn

2010

Jan. 15	Direct S	Capricorn
Feb. 10	Direct	Aquarius
Mar. 1	Direct	Pisces
Mar. 17	Direct	Aries

Apr. 2	Direct	Taurus
Apr. 18	Retrograde S	Taurus
May 12	Direct S	Taurus
June 10	Direct	Gemini
June 25	Direct	Cancer
July 9	Direct	Leo
July 28	Direct	Virgo
Aug. 21	Retrograde S	Virgo
Sept. 13	Direct S	Virgo
Oct. 3	Direct	Libra
Oct. 21	Direct	Scorpio
Nov. 9	Direct	Sagittarius
Dec. 1	Direct	Capricorn
Dec. 10	Retrograde S	Capricorn
Dec. 18	Retrograde	Sagittarius
Dec. 30	Direct S	Sagittarius

2011

Jan. 13	Direct	Capricorn
Feb. 4	Direct	Aquarius
Feb. 22	Direct	Pisces
Mar. 9	Direct	Aries
Mar. 31	Retrograde S	Aries
Apr. 23	Direct S	Aries
May 16	Direct	Taurus
June 3	Direct	Gemini
June 17	Direct	Cancer
July 2	Direct	Leo
July 28	Direct	Virgo
Aug. 3	Retrograde S	Virgo
Aug. 8	Retrograde	Leo
Aug. 27	Direct S	Leo
Sept. 9	Direct	Virgo
Sept. 26	Direct	Libra
Oct. 13	Direct	Scorpio
Nov. 2	Direct	Sagittarius
Nov. 24	Retrograde S	Sagittarius
Dec. 14	Direct S	Sagittarius

2012

Jan. 8	Direct	Capricorn
Jan. 27	Direct	Aquarius
Feb. 14	Direct	Pisces
Mar. 2	Direct	Aries
Mar. 12	Retrograde S	Aries
Mar. 23	Retrograde	Pisces
Apr. 4	Direct S	Pisces
Apr. 17	Direct	Aries
May 9	Direct	Taurus
May 24	Direct	Gemini
June 7	Direct	Cancer
June 26	Direct	Leo
July 15	Retrograde S	Leo
Aug. 8	Direct S	Leo
Sept. 1	Direct	Virgo
Sept. 17	Direct	Libra
Oct. 5	Direct	Scorpio
Oct. 29	Direct	Sagittarius
Nov. 7	Retrograde S	Sagittarius
Nov. 14	Retrograde	Scorpio
Nov. 27	Direct S	Scorpio
Dec. 11	Direct	Sagittarius
Dec. 31	Direct	Capricorn

2013

Jan. 19	Direct	Aquarius
Feb. 5	Direct	Pisces
Feb. 23	Retrograde S	Pisces
Mar. 18	Direct S	Pisces
Apr. 14	Direct	Aries
May 1	Direct	Taurus
May 16	Direct	Gemini
May 31	Direct	Cancer
June 26	Retrograde S	Cancer
July 20	Direct S	Cancer
Aug. 8	Direct	Leo
Aug. 24	Direct	Virgo

Sept. 9	Direct	Libra
Sept. 29	Direct	Scorpio
Oct. 21	Retrograde S	Scorpio
Nov. 11	Direct S	Scorpio
Dec. 5	Direct	Sagittarius
Dec. 24	Direct	Capricorn

2014

Jan. 12	Direct	Aquarius
Jan. 31	Direct	Pisces
Feb. 7	Retrograde S	Pisces
Feb. 13	Retrograde	Aquarius
Feb. 28	Direct S	Aquarius
Mar. 18	Direct	Pisces
Apr. 7	Direct	Aries
Apr. 23	Direct	Taurus
May 7	Direct	Gemini
May 29	Direct	Cancer
June 7	Retrograde S	Cancer
June 17	Retrograde	Gemini
July 1	Direct S	Gemini
July 13	Direct	Cancer
Aug. 1	Direct	Leo
Aug. 15	Direct	Virgo
Sept. 2	Direct	Libra
Sept. 28	Direct	Scorpio
Oct. 4	Retrograde S	Scorpio
Oct. 10	Retrograde	Libra
Oct. 26	Direct S	Libra
Nov. 9	Direct	Scorpio
Nov. 28	Direct	Sagittarius
Dec. 17	Direct	Capricorn

2015

Jan. 5	Direct	Aquarius
Jan. 21	Retrograde S	Aquarius
Feb. 11	Direct S	Aquarius
Mar. 13	Direct	Pisces

Mar. 31	Direct	Aries
Apr. 15	Direct	Taurus
May 1	Direct	Gemini
May 19	Retrograde S	Gemini
June 12	Direct S	Gemini
July 8	Direct	Cancer
July 23	Direct	Leo
Aug. 8	Direct	Virgo
Aug. 27	Direct	Libra
Sept. 17	Retrograde S	Libra
Oct. 9	Direct S	Libra
Nov. 2	Direct	Scorpio
Nov. 21	Direct	Sagittarius
Dec. 10	Direct	Capricorn

2016

Jan. 2	Direct	Aquarius
Jan. 5	Retrograde S	Aquarius
Jan. 9	Retrograde	Capricorn
Jan. 26	Direct S	Capricorn
Feb. 14	Direct	Aquarius
Mar. 5	Direct	Pisces
Mar. 22	Direct	Aries
Apr. 6	Direct	Taurus
Apr. 28	Retrograde S	Taurus
May 22	Direct S	Taurus
June 13	Direct	Gemini
June 30	Direct	Cancer
July 14	Direct	Leo
July 30	Direct	Virgo
Aug. 30	Retrograde S	Virgo
Sept. 22	Direct S	Virgo
Oct. 7	Direct	Libra
Oct. 25	Direct	Scorpio
Nov. 12	Direct	Sagittarius
Dec. 3	Direct	Capricorn
Dec. 19	Retrograde S	Capricorn

2017

Jan. 4	Retrograde	Sagittarius
Jan. 8	Direct S	Sagittarius
Jan. 12	Direct	Capricorn
Feb. 7	Direct	Aquarius
Feb. 26	Direct	Pisces
Mar. 14	Direct	Aries
Mar. 31	Direct	Taurus
Apr. 10	Retrograde S	Taurus
Apr. 20	Retrograde	Aries
May 3	Direct S	Aries
May 16	Direct	Taurus
June 7	Direct	Gemini
June 21	Direct	Cancer
July 6	Direct	Leo
July 26	Direct	Virgo
Aug. 13	Retrograde S	Virgo
Aug. 31	Retrograde	Leo
Sept. 5	Direct S	Leo
Sept. 10	Direct	Virgo
Sept. 30	Direct	Libra
Oct. 17	Direct	Scorpio
Nov. 6	Direct	Sagittarius
Dec. 3	Retrograde S	Sagittarius
Dec. 23	Direct S	Sagittarius

2018

Jan. 11	Direct	Capricorn
Jan. 31	Direct	Aquarius
Feb. 18	Direct	Pisces
Mar. 6	Direct	Aries
Mar. 23	Retrograde S	Aries
Apr. 15	Direct S	Aries
May 13	Direct	Taurus
May 30	Direct	Gemini
June 13	Direct	Cancer
June 29	Direct	Leo
July 26	Retrograde S	Leo

Aug. 19	Direct S	Leo
Sept. 6	Direct	Virgo
Sept. 22	Direct	Libra
Oct. 10	Direct	Scorpio
Oct. 31	Direct	Sagittarius
Nov. 17	Retrograde S	Sagittarius
Dec. 1	Retrograde	Scorpio
Dec. 7	Direct S	Scorpio
Dec. 13	Direct	Sagittarius

2019

Jan. 5	Direct	Capricorn
Jan. 24	Direct	Aquarius
Feb. 10	Direct	Pisces
Mar. 5	Retrograde S	Pisces
Mar. 28	Direct S	Pisces
Apr. 17	Direct	Aries
May 6	Direct	Taurus
May 21	Direct	Gemini
June 5	Direct	Cancer
June 27	Direct	Leo
July 8	Retrograde S	Leo
July 19	Retrograde	Cancer
Aug. 1	Direct S	Cancer
Aug. 12	Direct	Leo
Aug. 29	Direct	Virgo
Sept. 14	Direct	Libra
Oct. 3	Direct	Scorpio
Oct. 31	Retrograde S	Scorpio
Nov. 21	Direct S	Scorpio
Dec. 9	Direct	Sagittarius
Dec. 29	Direct	Capricorn

2020

Jan. 16	Direct	Aquarius
Feb. 3	Direct	Pisces
Feb. 17	Retrograde S	Pisces
Mar. 4	Retrograde	Aquarius

Mar. 10	Direct S	Aquarius
Mar. 16	Direct	Pisces
Apr. 11	Direct	Aries
Apr. 28	Direct	Taurus
May 12	Direct	Gemini
May 28	Direct	Cancer
June 18	Retrograde S	Cancer
July 12	Direct S	Cancer
Aug. 5	Direct	Leo
Aug. 20	Direct	Virgo
Sept. 6	Direct	Libra
Sept. 27	Direct	Scorpio
Oct. 14	Retrograde S	Scorpio
Oct. 28	Retrograde	Libra
Nov. 3	Direct S	Libra
Nov. 11	Direct	Scorpio
Dec. 2	Direct	Sagittarius
Dec. 21	Direct	Capricorn

2021

Jan. 8	Direct	Aquarius
Jan. 30	Retrograde S	Aquarius
Feb. 21	Direct S	Aquarius
Mar. 16	Direct	Pisces
Apr. 4	Direct	Aries
Apr. 19	Direct	Taurus
May 4	Direct	Gemini
May 30	Retrograde S	Gemini
June 23	Direct S	Gemini
July 12	Direct	Cancer
July 28	Direct	Leo
Aug. 12	Direct	Virgo
Aug. 30	Direct	Libra
Sept. 27	Retrograde S	Libra
Oct. 18	Direct S	Libra
Nov. 6	Direct	Scorpio
Nov. 24	Direct	Sagittarius
Dec. 13	Direct	Capricorn

2022

Jan. 2	Direct	Aquarius
Jan. 14	Retrograde S	Aquarius
Jan. 26	Retrograde	Capricorn
Feb. 4	Direct S	Capricorn
Feb. 15	Direct	Aquarius
Mar. 10	Direct	Pisces
Mar. 27	Direct	Aries
Apr. 11	Direct	Taurus
Apr. 30	Direct	Gemini
May 10	Retrograde S	Gemini
May 23	Retrograde	Taurus
June 3	Direct S	Taurus
June 13	Direct	Gemini
July 5	Direct	Cancer
July 19	Direct	Leo
Aug. 4	Direct	Virgo
Aug. 26	Direct	Libra
Sept. 10	Retrograde S	Libra
Sept. 23	Retrograde	Virgo
Oct. 2	Direct S	Virgo
Oct. 11	Direct	Libra
Oct. 30	Direct	Scorpio
Nov. 17	Direct	Sagittarius
Dec. 7	Direct	Capricorn
Dec. 29	Retrograde S	Capricorn

2023

Jan. 18	Direct S	Capricorn
Feb. 11	Direct	Aquarius
Mar. 3	Direct	Pisces
Mar. 19	Direct	Aries
Apr. 3	Direct	Taurus
Apr. 21	Retrograde S	Taurus
May 15	Direct S	Taurus
June 11	Direct	Gemini
June 27	Direct	Cancer
July 11	Direct	Leo

July 29	Direct	Virgo
Aug. 24	Retrograde S	Virgo
Sept. 16	Direct S	Virgo
Oct. 5	Direct	Libra
Oct. 22	Direct	Scorpio
Nov. 10	Direct	Sagittarius
Dec. 1	Direct	Capricorn
Dec. 13	Retrograde S	Capricorn
Dec. 23	Retrograde	Sagittarius

2024

Jan. 2	Direct S	Sagittarius
Jan. 14	Direct	Capricorn
Feb. 5	Direct	Aquarius
Feb. 23	Direct	Pisces
Mar. 10	Direct	Aries
Apr. 2	Retrograde S	Aries
Apr. 25	Direct S	Aries
May 15	Direct	Taurus
June 3	Direct	Gemini
June 17	Direct	Cancer
July 2	Direct	Leo
July 26	Direct	Virgo
Aug. 5	Retrograde S	Virgo
Aug. 15	Retrograde	Leo
Aug. 29	Direct S	Leo
Sept. 9	Direct	Virgo
Sept. 26	Direct	Libra
Oct. 14	Direct	Scorpio
Nov. 3	Direct	Sagittarius
Nov. 26	Retrograde S	Sagittarius
Dec. 16	Direct S	Sagittarius

2025

Jan. 8	Direct	Capricorn
Jan. 28	Direct	Aquarius
Feb. 14	Direct	Pisces
Mar. 3	Direct	Aries

Mar. 15	Retrograde S	Aries
Mar. 30	Retrograde	Pisces
Apr. 7	Direct S	Pisces
Apr. 16	Direct	Aries
May 10	Direct	Taurus
May 26	Direct	Gemini
June 9	Direct	Cancer
June 27	Direct	Leo
July 18	Retrograde S	Leo
Aug. 11	Direct S	Leo
Sept. 2	Direct	Virgo
Sept. 18	Direct	Libra
Oct. 6	Direct	Scorpio
Oct. 29	Direct	Sagittarius
Nov. 10	Retrograde S	Sagittarius
Nov. 19	Retrograde	Scorpio
Nov. 29	Direct S	Scorpio
Dec. 12	Direct	Sagittarius

2026

Jan. 2	Direct	Capricorn

APPENDIX D

Mercury Retrograde at a Glance
(Eastern Standard Time Zone)

The following tables show the dates for Mercury Retrograde for the fifty years beginning in 2000, as well as the sign in which Mercury is retrograde. In those instances where two signs are listed, such as "Scorpio/Libra," Mercury initially goes retrograde in the sign of Scorpio, then backs up into the sign of Libra before going direct again.

The tables were programmed by Rique Pottenger at ACS and were calculated for the Eastern Standard Time Zone. Thus the dates for Mercury Retrograde may vary by a day according to the different time zones around the world.

2000

Feb. 21–Mar. 15	Pisces
June 23–July 17	Cancer
Oct. 18–Nov. 8	Scorpio/Libra

2001

Feb. 4–Feb. 25	Pisces/Aquarius
June 4–June 28	Gemini
Oct. 2–Oct. 23	Libra

2002

Jan. 19–Feb. 8	Aquarius/Capricorn
May 15–June 8	Gemini
Sept. 15–Oct. 7	Libra/Virgo

2003

Jan. 2–Jan. 23	Capricorn
Apr. 26–May 20	Taurus
Aug. 28–Sept. 20	Virgo
Dec. 17–Jan. 6, 2004	Capricorn/Sagittarius

2004

Apr. 7–Apr. 30	Taurus/Aries
Aug. 10–Sept. 2	Virgo/Leo
Nov. 30–Dec. 20	Sagittarius

2005

Mar. 20–Apr. 12	Aries
July 23–Aug. 16	Leo
Nov. 14–Dec. 4	Sagittarius/Scorpio

2006

Mar. 3–Mar. 25	Pisces
July 5–July 29	Leo/Cancer
Oct. 29–Nov. 18	Scorpio

2007

Feb. 14–Mar. 8	Pisces/Aquarius
June 16–July 10	Cancer
Oct. 12–Nov. 2	Scorpio/Libra

2008

Jan. 29–Feb. 19	Aquarius
May 26–June 19	Gemini
Sept. 24–Oct. 16	Libra

2009

Jan. 11–Feb. 1	Aquarius/Capricorn
May 7–May 31	Gemini/Taurus
Sept. 7–Sept. 29	Libra/Virgo
Dec. 26–Jan. 15, 2010	Capricorn

2010

Apr. 18–May 12	Taurus
Aug. 21–Sept. 13	Virgo
Dec. 10–Dec. 30	Capricorn/Sagittarius

2011

Mar. 31–Apr. 23	Aries
Aug. 3–Aug. 27	Virgo/Leo
Nov. 24–Dec. 14	Sagittarius

2012

Mar. 12–Apr. 4	Aries/Pisces
July 15–Aug. 8	Leo
Nov. 7–Nov. 27	Sagittarius/Scorpio

2013

Feb. 23–Mar. 18	Pisces
June 26–July 20	Cancer
Oct. 21–Nov. 11	Scorpio

2014

Feb. 7–Feb. 28	Pisces/Aquarius
June 7–July 1	Cancer/Gemini
Oct. 4–Oct. 26	Scorpio/Libra

2015

Jan. 21–Feb. 11	Aquarius
May 19–June 12	Gemini
Sept. 17–Oct. 9	Libra

2016

Jan. 5–Jan. 26	Aquarius/Capricorn
Apr. 28–May 22	Taurus
Aug. 30–Sept. 22	Virgo
Dec. 19–Jan. 8, 2017	Capricorn/Sagittarius

2017

Apr. 10–May 3	Taurus/Aries
Aug. 13–Sept. 5	Virgo/Leo
Dec. 3–Dec. 23	Sagittarius

2018

Mar. 23–Apr. 15	Aries
July 26–Aug. 19	Leo
Nov. 17–Dec. 7	Sagittarius/Scorpio

2019

Mar. 5–Mar. 28	Pisces
July 8–Aug. 1	Leo/Cancer
Oct. 31–Nov. 21	Scorpio

2020

Feb. 17–Mar. 10	Pisces/Aquarius
June 18–July 12	Cancer
Oct. 14–Nov. 3	Scorpio/Libra

2021

Jan. 30–Feb. 21	Aquarius
May 30–June 23	Gemini
Sept. 27–Oct. 18	Libra

2022

Jan. 14–Feb. 4	Aquarius/Capricorn
May 10–June 3	Gemini/Taurus
Sept. 10–Oct. 2	Libra/Virgo
Dec. 29–Jan. 18, 2023	Capricorn

2023

Apr. 21–May 15	Taurus
Aug. 24–Sept. 16	Virgo
Dec. 13–Jan. 2, 2024	Capricorn/Sagittarius

2024

Apr. 2–Apr. 25	Aries
Aug. 5–Aug. 29	Virgo/Leo
Nov. 26–Dec. 16	Sagittarius

2025

Mar. 15–Apr. 7	Aries/Pisces
July 18–Aug. 11	Leo
Nov. 10–Nov. 29	Sagittarius/Scorpio

2026

Feb. 26–Mar. 21	Pisces
June 29–July 24	Cancer
Oct. 24–Nov. 13	Scorpio

2027

Feb. 9–Mar. 3	Pisces/Aquarius
June 10–July 5	Cancer/Gemini
Oct. 7–Oct. 28	Scorpio/Libra

2028

Jan. 24–Feb. 14	Aquarius
May 21–June 14	Gemini
Sept. 19–Oct. 11	Libra

2029

Jan. 7–Jan. 27	Aquarius/Capricorn
May 2–May 26	Taurus
Sept. 2–Sept. 25	Libra/Virgo
Dec. 22–Jan. 11, 2030	Capricorn

2030

Apr. 13–May 7	Taurus/Aries
Aug. 16–Sept. 8	Virgo
Dec. 6–Dec. 26	Capricorn/Sagittarius

2031

Mar. 26–Apr. 18	Aries
July 29–Aug. 22	Leo
Nov. 20–Dec. 9	Sagittarius/Scorpio

2032

Mar. 7–Mar. 30	Aries/Pisces
July 10–Aug. 3	Leo/Cancer
Nov. 2–Nov. 22	Sagittarius/Scorpio

2033

Feb. 19–Mar. 13	Pisces
June 21–July 15	Cancer
Oct. 17–Nov. 6	Scorpio/Libra

2034

Feb. 2–Feb. 24	Aquarius
June 2–June 26	Gemini
Sept. 30–Oct. 21	Libra

2035

Jan. 17–Feb. 7	Aquarius/Capricorn
May 13–June 6	Gemini/Taurus
Sept. 13–Oct. 5	Libra/Virgo

2036

Jan. 1–Jan. 21	Capricorn
Apr. 23–May 17	Taurus
Aug. 26–Sept. 17	Virgo
Dec. 15–Jan. 4, 2037	Capricorn/Sagittarius

2037

Apr. 5–Apr. 28	Taurus/Aries
Aug. 8–Sept. 1	Virgo/Leo
Nov. 29–Dec. 18	Sagittarius

2038

Mar. 18–Apr. 10	Aries/Pisces
July 21–Aug. 14	Leo
Nov. 12–Dec. 2	Sagittarius/Scorpio

2039

Mar. 1–Mar. 24	Pisces
July 3–July 27	Cancer
Oct. 27–Nov. 16	Scorpio

2040

Feb. 12–Mar. 5	Pisces/Aquarius
June 13–July 7	Cancer
Oct. 9–Oct. 30	Scorpio/Libra

2041

Jan. 26–Feb. 16	Aquarius
May 24–June 17	Gemini
Sept. 22–Oct. 14	Libra

2042

Jan. 10–Jan. 30	Aquarius/Capricorn
May 5–May 29	Taurus
Sept. 5–Sept. 28	Libra/Virgo
Dec. 25–Jan. 14, 2043	Capricorn

2043

Apr. 16–May 10	Taurus
Aug. 19–Sept. 11	Virgo
Dec. 9–Dec. 28	Capricorn/Sagittarius

2044

Mar. 28–Apr. 20	Aries
July 31–Aug. 24	Leo
Nov. 21–Dec. 11	Sagittarius

2045

Mar. 10–Apr. 2	Aries/Pisces
July 13–Aug. 6	Leo/Cancer
Nov. 5–Nov. 25	Sagittarius/Scorpio

2046

Feb. 21–Mar. 16	Pisces
June 24–July 19	Cancer
Oct. 19–Nov. 9	Scorpio

2047

Feb. 5–Feb. 27	Pisces/Aquarius
June 5–June 29	Cancer/Gemini
Oct. 3–Oct. 24	Scorpio/Libra

2048

Jan. 20–Feb. 10	Aquarius/Capricorn
May 16–June 9	Gemini
Sept. 15–Oct. 7	Libra/Virgo

2049

Jan. 3–Jan. 23	Capricorn
Apr. 26–May 20	Taurus
Aug. 29–Sept. 20	Virgo
Dec. 18–Jan. 6, 2050	Capricorn/Sagittarius

2050

Apr. 8–May 2	Taurus/Aries
Aug. 11–Sept. 3	Virgo/Leo
Dec. 1–Dec. 21	Sagittarius

BIBLIOGRAPHY

Bak, Per. *How Nature Works: The Science of Self-Organized Criticality.* New York: Copernicus/Springer Verlag, 1996.

Barton, Tamsyn. *Ancient Astrology.* London and New York: Routledge, 1994.

Blaze, Chrissie. *Mercury Retrograde: Your Survival Guide to Astrology's Most Precarious Time of Year!* New York: Warner, 2002.

Bryant, Rebecca. "Two-Timing the Clock." *Business 2.0,* February 2000.

Campbell, Joseph C. *The Inner Reaches of Outer Space: Metaphor as Myth and as Religion.* New York: Harper & Row, 1986.

Carotenuto, Aldo. *Eros and Pathos: Shades of Love and Suffering.* Toronto: Inner City Books, 1987.

Carter, Stephen. *Integrity.* New York: Basic Books, 1996.

Clark, Brian. "Gemini: Searching for the Missing Twin." Cedar Ridge, Calif.: the *Mountain Astrologer,* June 2000.

Coelho, Paulo. *The Alchemist.* London: Thorsons/HarperCollins, 1995.

Conforti, Michael. *Field, Form, and Fate: Patterns in Mind, Nature, and Psyche.* Woodstock, Conn.: Spring Publications, 1999.

Cumont, Franz. *Astrology and Religion among the Greeks and Romans (1912).* Kessinger Publishing, www.kessinger.net.

Edis, Freda. *The God Between: A Study of the Astrological Mercury.* London/New York: Penguin/Arkana, 1995.

Fagles, Robert, trans. *The Odyssey of Homer.* Introduction and notes by Bernard Knox. New York: Viking, 1996.

Fitzgerald, Robert, trans. *The Iliad of Homer.* New York: Anchor Press/Doubleday, 1974.

France, Peter. *Hermits: The Insights of Solitude.* New York: St. Martin's Press, 1996.

Francis, Eric. "Spicing Up Your Mercury Retrograde." www.stariq.com., July 6, 2000.

Gerhardt, Dana. "Zodiac Tales: A Thousand and One Nights of Gemini." Cedar Ridge, Calif.: *The Mountain Astrologer,* October/November 1999.

George, Demetra. "A Golden Thread: The Cultural Transmission of Astrology." Cedar Ridge, Calif.: *The Mountain Astrologer,* August/September 2003.

Grafton, Anthony. *Cardano's Cosmos: The Worlds and Work of a Renaissance Astrologer.* Boston: Harvard University Press, 2001.

Greene, Liz, and Howard Sasportas. *The Inner Planets: Building Blocks of Personal Reality.* York Beach, Me.: Samuel Weiser, 1993.

Hand, Robert. *Planets in Transit: Life Cycles for Living.* Westchester, Pa.: Whitford Press/Schiffer Publishing, 1976.

Hannon, Geraldine Hatch. "Interview with Physicist-Astrologer Will Keepin." Cedar Ridge, Calif: *The Mountain Astrologer,* February/March 1997.

Harrison, Tony, and Rex Collings, trans. Aeschylus' Choephoni, in *The Oresteia,* New York: Viking Press, 1982.

Hickey, Isabel M. *Astrology, A Cosmic Science.* USA, Canada, and UK: CRCS Publications, 1992.

Hillman, James. *Healing Fiction.* Dallas: Spring Publications, 1983.

———. *The Archetypal Phenomenon of Astrology.* February 10, 1997, lecture presented at Psyche and Symbols Conference. Available on tape from www.isisinstitute.com.

Jankowiak, William, ed. *Romantic Passion: A Universal Experience?* New York: Columbia University Press, 1995.

Jung, C. G. *Memories, Dreams, Reflections.* New York: Vintage Books, 1965.

———. "The Psychology of the Trickster Archetype." *The Archetypes and the Collective Unconscious,* vol. 9. Princeton, New Jersey: Princeton University Press.

———. "Commentary by C. G. Jung." *The Secret of the Golden Flower: A Chinese Book of Life,* translated by Richard Wilhelm. New York: Harcourt, Brace & World, 1962.

Katzanzantzakis, Nikos. *Zorba the Greek.* New York: Simon & Schuster, 1952.

Kerenyi, Karl. *Hermes, Guide of Souls: The Mythologem of the Masculine Source of Life.* Dallas: Spring Publications, 1986.

Khan, Hazrat Inayat. *Tales: Told by Hazrat Inayat Khan.* New Lebanon, New York: Sufi Order Publications, 1980.

———. *The Inner Life.* Boston: Shambhala Publications, 1997.

Khan, Pir Vilayat Inayat. *Awakening: A Sufi Experience.* New York: Tarcher/Putnam, 1999.

Krishnamurti, J. *This Light in Oneself: True Meditation.* Boston: Shambhala Publications, 1999.

Lattimore, Richmond, trans. *The Iliad of Homer.* Chicago and London: University of Chicago Press, 1951.

Levine, Rick. "Mercury Retrograde: A Modern Look." www.stariq.com, Nov. 8, 1999.

Lonsdale, Elias. *Inside Planets.* Berkeley, Calif.: North Atlantic Books, 1999.

Middlebrook, Diane. *Her Husband: Hughes and Plath—A Marriage.* New York: Viking, 2003.

Moore, Thomas. Foreword. *Timeshifting: Creating More Time to Enjoy Your Life.* New York: Doubleday, 1996.

Otto, Walter F. *The Homeric Gods: The Spiritual Significance of Greek Religion.* London and New York: Thames and Hudson/Pantheon Books, 1979.

Pagels, Elaine. *The Origin of Satan.* New York: Vintage Books, 1996.

Paris, Ginette C. *Pagan Grace: Dionysos, Hermes and Goddess Memory in Daily Life.* Woodstock, Conn.: Spring Publications, 1990.

Rechstchaffen, Stephan, M.D. *Timeshifting: Creating More Time to Enjoy Your Life.* New York: Doubleday, 1996.

Rifkin, Jeremy. *Time Wars: The Primary Conflict in Human History.* Henry Holt and Company, 1987.

Rothstein, Edward. "An Astrological Quirk Brings Fear and Trembling to Some Corners of the Internet." *New York Times,* 30 Sept. 1996.

Sakoian, Frances B., and Louis S. Acker. *The Importance of Mercury in the Horoscope.* Frances Sakoian, 1973. (Note: this is a self-published pamphlet.)

Schmidt, Robert. *The Problem of Astrology.* www.projecthindsight.com.

Schulman, Martin. *Karmic Astrology: The Moon's Nodes and Reincarnation.* York Beach, Me.: Samuel Weiser, 1989.

Seymour, Percy, Ph.D. "Astrologers by Nature." Cedar Ridge, Calif: *The Mountain Astrologer,* February/March 2002.

Sholly, Kate. "Interview with Richard Tarnas on Transforming the Western World View." Cedar Ridge, Calif.: *The Mountain Astrologer,* December 1995.

Sluyter, Dean. *The Zen Commandments: Ten Suggestions For a Life of Inner Freedom.* New York: Tarcher/Penguin Putnam, 2001.

Smith, Huston. *The Religions of Man.* New York: Harper and Row, 1958.

Star, Jonathan, trans. *Rumi: In the Arms of the Beloved.* New York: Tarcher/Penguin, 1997.

Stein, Murray. *In MidLife.* Dallas: Spring Publications, 1983.

Sullivan, Erin. *Retrograde Planets: Traversing the Inner Landscape.* York Beach, Me.: Samuel P. Weiser, 1992.

Swerdlow, N. M., ed. *Ancient Astronomy and Celestial Divination.* London: Massachusetts Institute of Technology, 1999.

Tarnas, Richard. *The Passion of the Western Mind: Understanding the Ideas That Have Shaped Our World.* New York: Harmony Books, 1991.

Terriktar, Tem. "Mercury Retrograde Strikes Again!" Cedar Ridge, Calif.: *The Mountain Astrologer,* February/March 2001.

Van De Castle, Robert. L., Ph.D. *Our Dreaming Mind: A Sweeping Exploration of the Role That Dreams Have Played in Politics, Art, Religion, and Psychology, from Ancient Civilizations to the Present Day.* New York: Ballantine, 1994.

Van Dyke, Henry. *The Story of the Other Wise Man.* New York: Ballantine/Epiphany Books, 1984.

Von Franz, Marie-Louise. *Creation Myths.* Dallas: Spring Publications, 1972.

Warnock, Christopher, Esq. "Of Mercury, His Signification, Nature and Property." From William Lilly's *Christian Astrology* (1st edition, 1647, reprinted 1985 Regulus). www.renaissanceastrology.com.

Walker, Barbara G. *The Woman's Encyclopedia of Myths and Secrets.* New York: HarperCollins, 1983.

Wheelis, Allen. *How People Change.* New York: HarperPerennial, 1973.

Whitfield, Peter. *Astrology: A History.* New York: Harry N. Abrams, 2001.

Wilkinson, Robert. *A New Look at Mercury Retrograde.* York Beach, Me.: Samuel Weiser, 1997.

Wilson, Edward O. *Consilience: Toward a Unity of Knowledge.* New York: Vintage Books, 1998.

INDEX

ABOUT THE AUTHOR

Pythia Peay has been writing, teaching, and studying spiritual themes for more than thirty years. Her writing on a wide range of spiritual subjects has appeared in newspapers and magazines around the country. She is a contributor to Religion News Service, and her meditations and articles appear regularly on the multifaith Web site Beliefnet.com. She is a member of the Sacred Circles committee on women's spirituality at the Washington National Cathedral, as well as the Center for Contemplative Mind in Society. She teaches Sufi meditation classes, and she has collaborated with the teacher Pir Vilayat Inayat Khan on his book *Awakening: A Sufi Experience.* The author of *Soul Sisters: The Five Sacred Qualities of a Woman's Soul,* Peay offers workshops on the Divine Feminine, and is also a practicing astrologer. She is the mother of three grown sons, and lives in the Washington, D.C., area. Her Web site is *www.pythiapeay.com.*